D1011238

Making Sense of Weather and Climate

MARK DENNY

Making Sense of Weather and Climate

The Science Behind the Forecasts

COLUMBIA UNIVERSITY PRESS NEW YORK

Columbia University Press
Publishers Since 1893
New York Chichester, West Sussex
cup.columbia.edu

ISBN 978-0-231-17492-3 (cloth : alk. paper)
ISBN 978-0-231-54286-9 (e-book)

Cataloging-in-Publication Data is on
file at the Library of Congress.

∞

Columbia University Press books are printed on permanent
and durable acid-free paper.
Printed in the United States of America

COVER DESIGN: Diane Luger
COVER IMAGE: Chemical glass flask
(© Dollar Photo Club); Olexander, Sunset tornado (© Dollar Photo Club / James Thew)

Contents

Author's Note

One of the issues that arises when writing popular science books is that of units. Yards or meters? Kilograms or pounds? In everyday life, we use both. However, mixing units is considered uncool in the scientific community, though it happens often enough, and so in "Forecast," I offer a sheepish apology for this misdemeanor. (In fact, I use metric units with more familiar units added parenthetically, where they are needed.)[1]

References are a mixture of primary sources, which can be very technical, and secondary or educational material, which may be more suitable for the nonspecialist. The former are included mostly to provide backup for the claims made in the text, and for readers who want to dig much deeper into weather and climate physics; the latter are further reading for those who wish to delve only a little deeper into our subject. The notes point you to the references but do much more than that, so please read them.

Reading over the manuscript, I note quite a few statements of the kind "As we will see in chapter X" and "As we saw earlier." I appreciate that such temporal cross-referencing can be a little irritating to some readers, but it is unavoidable in subjects that are as intertwined and multifaceted as meteorology and climatology. Similarly, there are subjects—the Coriolis force is one—that are raised more than once, in different chapters. This is also due, in part at least, to the interconnectedness of matters meteorological and, in part, as a pedagogical device.

Acknowledgments

Writing this book has occupied something like a year of my working life. It required much organizing and reorganizing, explaining and rewriting, reviewing and reviewing again. I am grateful to Patrick Fitzgerald and Ryan Groendyk of Columbia University Press for their encouragement and support during this intense process. Thanks to Irene Pavitt, at Columbia University Press, and to Terry Kornak for turning the manuscript into a book. For his technical help, I am grateful to meteorologist Chris Wamsley of the National Oceanic and Atmospheric Administration. Finally, I thank Thomas Birner of Colorado State University for his very detailed, constructive, and helpful review of the first version of the manuscript. Any errors that remain are mine alone.

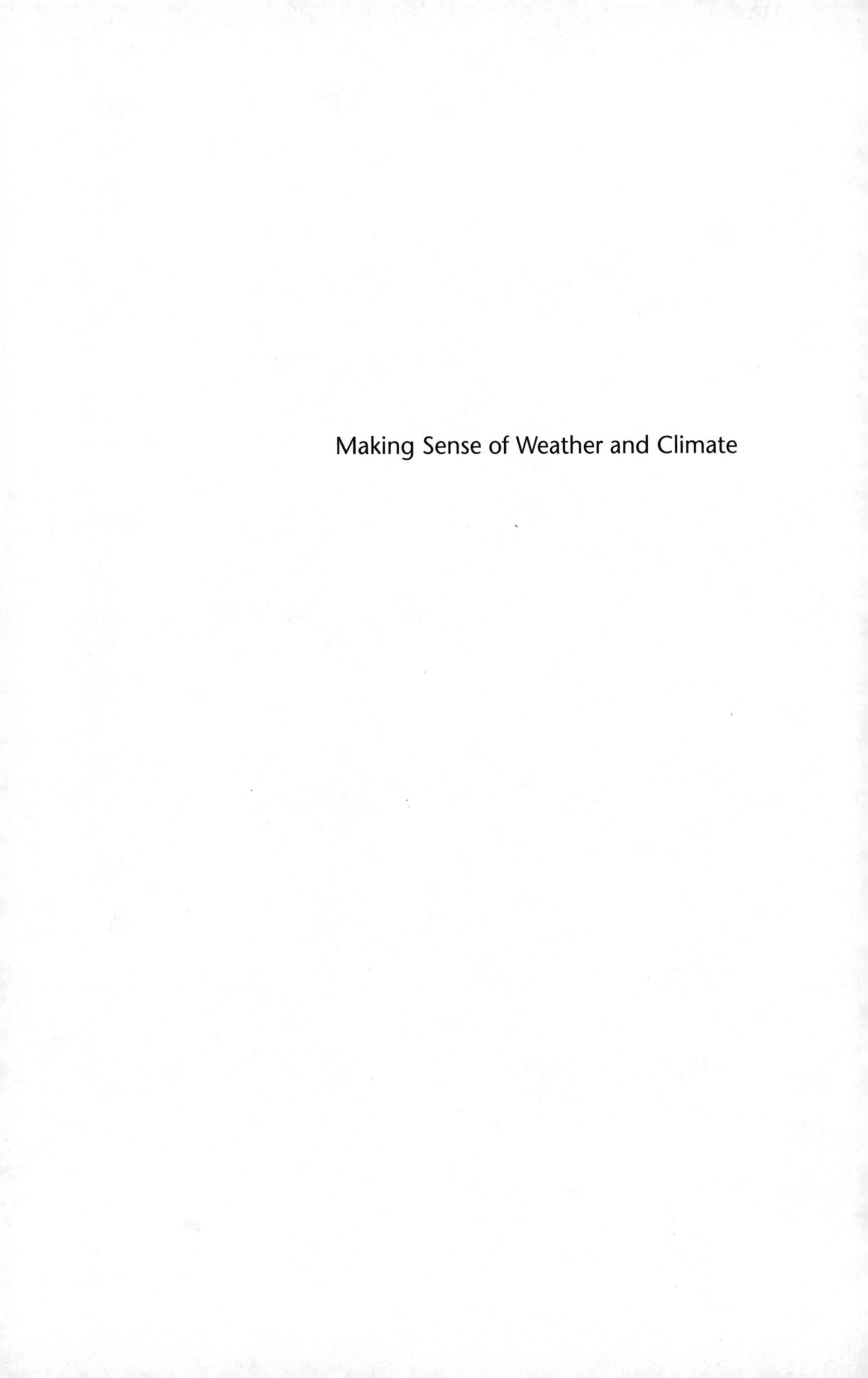

Making Sense of Weather and Climate

Forecast

It's so dry the trees are bribing the dogs.

Charles Martin

The book that you hold in your hands will help you make sense of our weather. My aim in writing such a book is, as always, to provide transparent science that conveys the core ideas underpinning a complex physical system—in this case, the physics of our atmosphere. This book will also help you make sense of our climate (so extending the physics to our oceans) and how we model and predict climate and influence climate change. Loosely, climate is average weather, and thus short-term climate prediction is a walk in the park compared with weather prediction. It is much easier to predict the average world temperature for next year, for example, than to predict the temperature at the bottom of your garden tomorrow morning. Even so, long-term climate predictions are far from easy, as we will see.[1]

Why did it rain today when the forecast said sun? More generally, why is it so hard for meteorologists to predict detailed weather accurately when the likely weather is so trivial to predict? ("The weather tomorrow will be the same as it is today" works about 70% of the time on average, depending on latitude.) Here are a few more questions that you may have asked yourself whenever the state of the weather intrudes on your thoughts. Why do thunderstorms happen? Weather is seasonal for obvious reasons, so why is the weather on my birthday not always the same? What is a weather front? If weather is so variable, why doesn't the temperature ever go to –100° or +200°? Why are there weather patterns, and will they be the same when my grandchildren grow up? Why can I see through rain but not fog?

Climate questions may intrude at a different level—they are perhaps more important but less urgent. Why, like the atmosphere, is the global-warming debate increasingly heated? Is human industry responsible for this global warming, and, if so, can we reverse the process? How much of climate change is natural and inevitable, whatever we do?

I answer these questions and many others in the chapters to follow. The explanations will be accessible to an intelligent reader with no background in meteorology or climatology, though the underlying physics is immensely complex. Concerning meteorology, I concentrate mostly on the weather you get in your backyard, not so much on the extremes that occasionally wreak havoc around the world.[2] (You will, however, finish this book with a working knowledge of tornadoes, hurricanes, thunderstorms, droughts, and floods.) Each topic is presented with readable prose but no hype or agenda. Metaphorically, your forecast is for sunshine, not moonshine.[3] One prosaic difference between weather and climate is politics, and politics does not mix well with rational debate or dispassionate analysis. So I will go to some effort to present you with the facts ("Just the facts, ma'am," as Joe Friday would say) and eliminate the politics.

Except that this is impossible, when the subject is climate change. I adopt the view of most (almost all) scientists that the accumulated data point to a changing climate and to human activity as the likely cause. The problem is that, merely by accepting these data and making the inference, I am declaring a political viewpoint even though I infer via rational scientific arguments and not from any political preconceptions. So be it: I follow where science leads, and if I end up in a political camp, then, from my perspective, I got there accidentally. Please try to do the same; readers will get more out of this book if they consider it to be what the writer considers it to be—an explanation of scientific phenomena, without any other agenda.

This is a good place to point out the position I hope that this book will occupy in the literature of popular weather and climate science.

Few nonspecialist books attempt to explain both weather and climate, and many of those that do are rather superficial. They are, to quote the physicist David Derbes, "a mile wide and a millimeter deep."[4] Here we attain the breadth of coverage while penetrating deeper into key aspects of our subject. Coverage cannot be comprehensive but will be deep enough to give insight.

Our subject is inherently statistical, and most humans have poor intuition about statistical matters. Even though I am a scientist who is used to statistical data, it sounds odd to me when I hear the weather presenter talk about the highest temperature of the day as an average ("Today's high was average for this time of year"), but it makes good sense. In chapter 6, I lead off my account of weather with a (light and readable, I hope) discussion of statistics and chaos in meteorology. Statistics as a subject may be drier than a summer day in Phoenix, but it doesn't have to be presented that way, and it is essential to any insightful understanding of the subject.

"Lead off" in chapter 6? Yes, indeed. The first three chapters set the table for the feast, by providing necessary background about the basic weather-generating mechanism of heat transfer (chapter 1), about the star we circle and the planet we live on (chapter 2), and about the atmosphere we live under (chapter 3). Chapter 4 looks into the slow, dynamical effects that drive climate change without influencing day-to-day weather. Chapter 5 tees up the statistics of chapter 6 (which you think you will hate, but you're wrong about that) by conveying to you the welter of data that feed our computer models of weather and climate. The remaining chapters build on these early foundations (our physical world is a complex system that takes a while to describe in a sensible way) and show you what we understand about weather and climate modeling and prediction.

I am not writing a textbook—there are already many meteorology and climatology texts out there. I am writing a solid account of weather physics and its slower sibling, climatology, that is aimed at the intelligent nonspecialist. You will need no more than high-school physics and math; there is a technical appendix with more math details for those of you who crave that sort of thing, but the main text is stand-alone. I want this book to be a breath of fresh air, not long winded. My approach is to explain key features of weather and climate physics with words and diagrams that get across the core ideas, backed up by more detailed examples set apart in boxes. One such example: hurricanes are an interesting and relevant phenomenon illustrating the Coriolis force, angular momentum, latent heat, and atmospheric instability, and so they merit such treatment.

Weather interests people for many reasons. It is one of the most relevant applications of science, affecting our daily lives. We don't tune into local radio or television every morning to learn about the Higgs boson or to see the latest discoveries about slime molds, but we want to know what the weather is going to do today. Maybe we want to know this information simply so that we can decide what to wear, but likely there are more important reasons. Will there be ice on the road to work? Will the smog be bad? Transportation, agriculture, health care, safety, military operations—all are affected by weather. Thus freezing rain may bring out highway-maintenance trucks dispensing salt and grit, while flooding or fog may lead to traffic diversions. Crops may be watered (covered) if a dry spell (frost) is forecast. A heat wave can be fatal to unprepared retirees,[5] while a storm at sea can be fatal to fishermen, and a tornado or wildfire fatal to anybody in its path. Military planners need to know about upcoming weather conditions in their area of operation (think of D-Day, or of air strikes in Bosnia). Taking the long view, if climate warming is real—and you will see that it is very real—then we need to know what its consequences will be. One of the important consequences is more extreme weather.

Our knowledge of weather physics, and especially our ability to predict weather, has greatly improved in recent decades; this is the subject of chapter 10. This trend will continue, though we will never be able to predict local weather accurately a month in advance (for reasons made crystal clear in chapter 6). My aims and hopes are twofold. First, having read this book, you will gain significant insight into the phenomena of weather and climate. Second, having read this book, you will better appreciate the considerable effort that is required to bring you the daily weather forecast.

1

Feeling the Heat

The sun, with all those planets revolving around it and dependent on it, can still ripen a bunch of grapes as if it had nothing else in the universe to do.

Galileo Galilei

Weather is all about air and water being moved around (and being heated and cooled). Such activity requires energy to drive it; the energy that drives our weather comes from the sun in the form of heat. In this chapter, we describe the basic physics that determines the energy balance of our planet and, in particular, will see why Earth's average temperature is what it is.

Local Astronomy

Galaxies are collections of billions of large thermonuclear reactors that we call "stars," loosely held together by gravity. Our local thermonuclear reactor, named "sun," is a ball of plasma tightly held together by gravity. The complicated nuclear reactions and the equally complicated fluid dynamics of our sun are not the subject of this book: suffice it to say that the sun is a rotating sphere of radius 696,000 kilometers (432,474 miles) with a surface temperature of about 5,500°C (9,900°F). Each square meter of this surface radiates

63 megawatts of electromagnetic power—that's the output of a small power station. For a refresher on electromagnetic (EM) radiation, see box 1.1. Do the math, and you will find that the total power generated by the sun (its *luminosity*), and radiated out to the rest of the universe, is about 3.85×10^{26} watts (385 trillion trillion watts).

The sun's tremendous power radiates out in all directions equally, and so the fraction of this power that bathes Earth is easy to work out (for interested readers, the simple geometric calculation is provided in the appendix). The result is that 1.7×10^{17} watts of solar EM radiation reaches our upper atmosphere.

There are a couple of important details that this simple estimate sweeps under the carpet, however. First it assumes that Earth's orbit about the sun is circular. This is not quite true: the orbit is elliptical—the closest point to the sun and the farthest point differ by 1.6%. This difference does not matter much for us; it is a good approximation to say that the ellipse is almost a circle. Second, the solar radiation that reaches our upper atmosphere is not the same as the radiation that heats Earth and drives its weather. The radiation absorbed by Earth is less than the incident radiation by about one-third as a result of reflection off the atmosphere, off clouds within it, and off the surface of our planet. We will investigate the details later; the idea is illustrated in figure 1.1. It is enough for now to say that the solar power that is absorbed by Earth is a (nearly) constant 1.23×10^{17} watts.

The output of our sun is not quite constant. As the sun is technically a variable star, its power oscillates slightly, varying in amplitude by about 1 W m^{-2} (watt per square meter) between the maximum and minimum values. This wobble is about 0.1% of the average power and so may seem to be a mere detail of interest only to astronomers, but in fact it has measurable consequences for our climate on Earth. The oscillations cycle every 11 years and correlate strongly with sunspot activity. Sunspots are magnetic disturbances that bubble to the surface. They are associated with polarity flips in the sun's magnetic field—the north and south poles of the sun flip over every 11 years.[1] There are other cycles that have been measured, either directly (from solar power output or magnetic field strengths) or indirectly (via proxy measurements on Earth, such as tree-ring growth and ^{14}C abundances), with periods of 80 years, 200 years, and 1,000 years. The solar power output variations resulting from these cycles lead to global temperature

Box 1.1
Electromagnetic Radiation

Light and microwaves; ultraviolet radiation and heat (also known as "infrared radiation"); and radio waves, X-rays, and gamma rays are all the same in that they are EM waves that all travel at the universal speed limit—the speed of light. They differ only in frequency (equivalently, in wavelength) and in the amount of energy they carry, which is proportional to frequency. Thus a ray of ultraviolet light of frequency 10^{16} Hz (10,000,000,000,000,000 cycles per second) has 100 times as much energy as an infrared ray of frequency 10^{14} Hz. Box figure 1.1 shows a broad section of the EM spectrum. What we call "light" is that small band that our eyes can detect. What we call "heat" is a lower frequency band that we can feel but not see. The sun's energy is mostly EM radiation at infrared, visible, and ultraviolet frequencies.

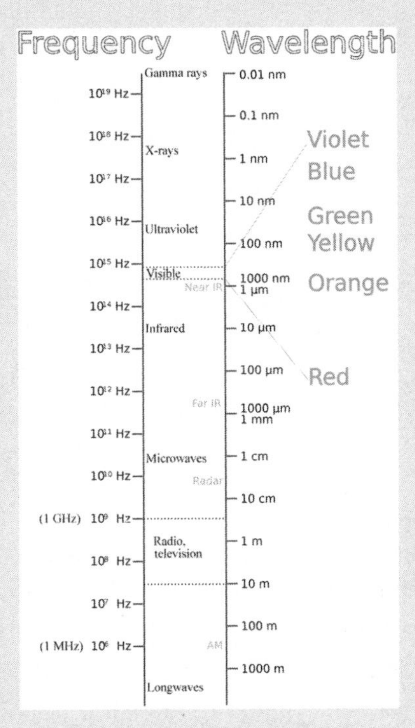

Box Figure 1.1 The electromagnetic spectrum. Frequency is expressed in Hz (cycles per second), and wavelength is in meters (m), centimeters (cm), millimeters (mm), micrometers (μm [millionths of a meter]), and nanometers (nm [billionths of a meter]). The energy of a photon (an elementary particle of light, zillions of which constitute an electromagnetic wave) is proportional to its frequency. The electromagnetic power emitted by the sun peaks in the visible region. (Adapted by the author from a figure by Victor Blacus)

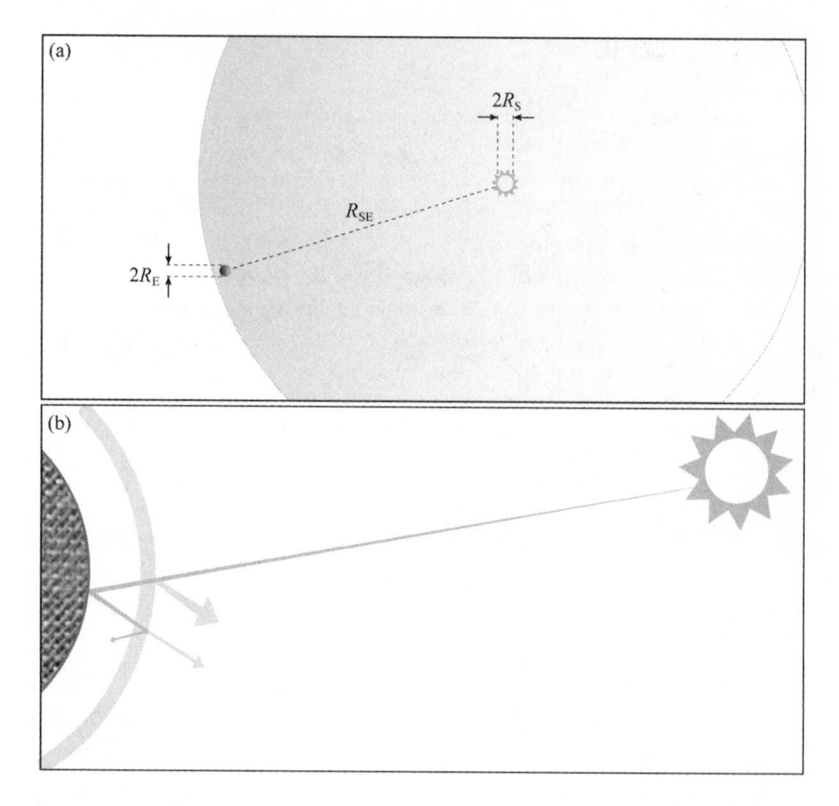

Figure 1.1 Incoming solar power. (*a*) The orbit of Earth is approximately circular, of radius 150 million kilometers (93 million miles) (R_{SE}). Electromagnetic power from the sun (radius R_S) spreads evenly in all directions; given the size of Earth (radius R_E) and its distance from the sun, it is not hard to work out the solar power that is intercepted by our planet (the calculation is in the appendix). (*b*) A ray of sunlight is either absorbed at the surface or reflected off it, or is absorbed by the atmosphere or reflected off it. About one-third of incident electromagnetic radiation is reflected back out into space—mostly from clouds. The rest is absorbed.

variations here on Earth. This connection has been established for several hundred years dating from the Middle Ages to the present day. Thus the Spörer minimum (1410–1540) and the Maunder minimum (1645–1715)—years of harsh winters and cool and wet summers—correspond to periods of low solar activity. A strong statistical correlation between solar power and global mean surface temperatures since 1610 has been established recently (though this correlation certainly does not account for global warming).[2]

The Blue-Green Planet

In chapter 2, we will need more detail, about the differing characteristics of Earth's surface—for example, how land and ocean reflect EM radiation to differing degrees. Here we are concentrating on the amount of heat energy from the sun that is absorbed by our planet. What about that other source of heat, known to every underground miner? Near the surface, for every kilometer (0.6 mile) that we descend underground, the temperature increases by 25°C (45°F). The center of Earth is believed to be solid iron at about 7,000°C (12,600°F), under very high pressure. This inner core is surrounded by a fluid outer core consisting mostly of molten iron. Between the outer core and the surface crust is the mantle, an inhomogeneous mixture of hot or molten rock. Where does all this heat come from, how much is there, and how much does it contribute to our weather and climate?

It was once thought that the internal heat of Earth was left over from a primordial fireball. That is, our planet began life as a molten blob of material—rock and metal—that coalesced and cooled. Indeed, at the end of the nineteenth century, the eminent physicist William Thomson (elevated to a peerage as Lord Kelvin)[3] estimated the age of our planet from the rate at which the molten blob cooled. He came up with various figures by this method—100 million years, 24 million, 20 million—which pleased no one. At the time many people believed that Earth was created on October 22 or 23, 4004 B.C.E., according to calculations based on a literal reading of the Bible made by an Irish archbishop, James Ussher, in 1650. But the new science of geology was insisting on a much older age for our planet—many hundreds of millions of years. During this period, the relative roles of science and religion were frequently debated, in public as well as in the technical literature. (Evolutionary theory had been proposed by Charles Darwin a few decades earlier, and was very controversial at the time. The evolutionists had a dog in the "age of Earth" fight; they sided with the geologists in advocating a much older planet than Kelvin calculated.) Some of these debates became very heated—an appropriate description given Kelvin's calculation.

The resolution came from an up-and-coming physicist named Ernest Rutherford, a New Zealander working at Cambridge University who, in an address in 1904 to a large audience that included Lord Kelvin, showed that radioactivity changed the game. Rutherford was a leading light in the new generation of scientists studying what we now call nuclear physics. Radioactivity had been discovered at the end of the nineteenth century by Henri Becquerel in France. Rutherford and his colleagues knew that many radioactive substances existed naturally in Earth's interior and that these substances generate heat. Later calculations and measurements showed that radioactivity generates heat within our planet at the rate of 30 terawatts (1 terawatt is 1 trillion watts), which is more than the present-day power consumption of humanity. Also, measurements of radioactive decay show that Earth is 4,540 ± 50 million years old, in line with the requirements of geology and evolutionary biology.[4]

Thirty terawatts is a huge amount of power, but it is a drop in the bucket compared with the power absorbed by our planet from the sun. Solar heating contributes 4,000 times as much heat to Earth's energy budget than does internal heating due to radioactivity. Thus we can answer the second question posed at the start of this section and say that the internal heat of our planet contributes negligibly to its heating and so contributes negligibly to our weather and climate. To a very good approximation, we can say that weather and climate arise from heating of Earth's surface by the sun.

Blackbody Radiation

To understand the heating of our atmosphere and planetary surface by the sun in more detail, we need to appreciate an important concept in thermodynamics, that of a *blackbody*.[5] This idealized object is a perfect absorber of heat; it reflects nothing and so would appear to be perfectly black, like soot—and soot is in fact pretty close to being an ideal blackbody. Such an idealization does not exist in the real world; heat is reflected off the surface, to a greater or lesser extent (as we have seen for Earth reflecting solar radiation). Even soot reflects some light; after all, we can see it. Nevertheless, the blackbody concept is a useful one for physicists because it *almost* applies in many cases—it is a good approximation—and because it can be analyzed theoretically.

A star such as our sun is very nearly a blackbody that is in thermal equilibrium with its surroundings; that is, it has a more or less constant surface temperature. A stove, similarly, is approximately a blackbody at thermal equilibrium.[6] A blackbody that is in thermal equilibrium with its surroundings has a characteristic emission spectrum that is a function of its surface temperature. (This equation is discussed in the appendix.) That is, the amount of EM radiation it emits at each given frequency can be calculated. The power transmitted turns out to depend on only the surface temperature. This spectrum is shown for a blackbody with the same surface temperature as the sun in figure 1.2. Also shown in the figure is the actual spectrum of EM power of our sun—note that it follows closely the blackbody spectrum.

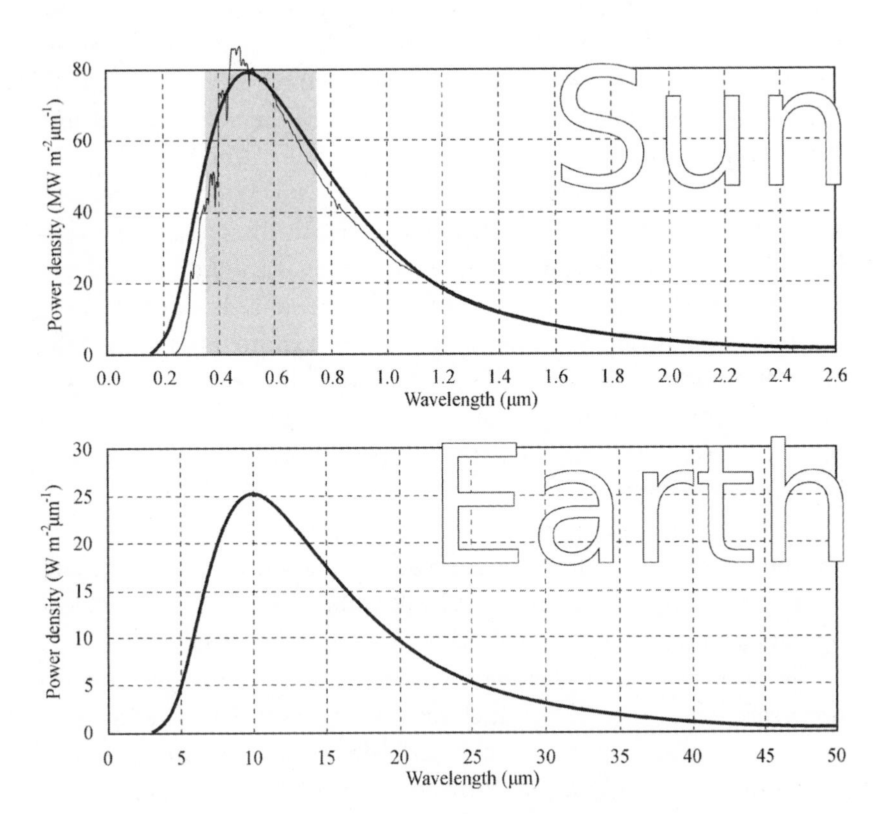

Figure 1.2 Emission spectra (power density per unit wavelength) for blackbodies with the same surface temperature as the sun and for Earth (*thick lines*). The actual solar spectrum (*thin line in upper graph*) is much more ragged, due to details of solar nuclear reactions and solar composition, but closely follows the blackbody curve. The visible part of the spectrum is shaded. Note that for Earth, the peak of the emission spectrum is at about 10 micrometers, in the infrared region. Note also the different scales of the two graphs.

Note that the peak of the solar spectrum is in the visible range; presumably that is why our eyes evolved to see these frequencies. Note also that almost all the power is in the band of frequencies that are visible or at somewhat lower frequencies (that is, somewhat longer wavelengths [infrared radiation, to the right of the visible band in figure 1.2]). A little of the power is at somewhat higher frequencies (shorter wavelengths [ultraviolet radiation, to the left of the visible band]). I will refer to the region of the EM spectrum that contains most of the power emitted by the sun as *shortwave radiation*.

Earth has an average surface temperature of 15°C (59°F). Our planet is also a blackbody (nearly) that is in thermal equilibrium with its surroundings, and so should it have an EM spectrum (given that, so far, I have said only that it *absorbs* radiation from the sun)? Our planet has an average surface temperature that is nearly constant—it does not change much over time, or changes only slowly—and so we can say that it is in thermal equilibrium with its surroundings. It absorbs heat from the sun every hour of every day, year in and year out, in the form of shortwave radiation; yet its temperature is constant, so it must also be emitting heat. The amount of heat emitted must equal the amount absorbed; otherwise, the planet would heat up or cool down. So Earth emits blackbody radiation, with a spectrum as shown in figure 1.2. Note that Earth absorbs shortwave radiation but emits heat in the infrared region, henceforth dubbed *longwave radiation*.

This difference is crucial for what follows, so let me repeat it. Earth absorbs energy in the form of shortwave radiation from the sun and reemits it as longwave radiation. Why does this difference in frequency/wavelength matter? It matters because it leads to the greenhouse effect.

The thermodynamic treatment of blackbody radiation gives rise to the *Stefan–Boltzmann Law*, which relates the power density at the surface of a blackbody to its surface temperature. This law can be applied to the sun and to Earth, to predict what the average temperature of Earth should be given the amount of EM power that it absorbs from the sun. In the appendix, we show that such a calculation leads to a mean (average) surface temperature for our planet of –19°C (–2.2°F). We must understand the difference.

Other planets in our solar system are also close to being blackbodies, and so we can calculate what their surface temperatures ought to be. The results are shown in figure 1.3 for the four inner planets. Blackbody

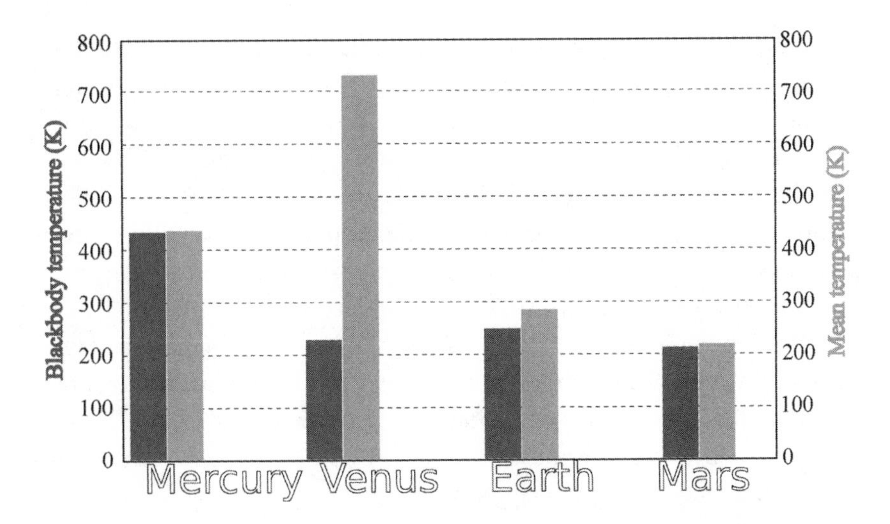

Figure 1.3 Calculated blackbody temperatures of the four inner planets and their actual mean surface temperatures (absolute temperature scale). (Data from American Chemical Society, "Energy from the Sun," ACS: Chemistry for Life, https://www.acs.org/content/acs/en/climatescience/energybalance/energyfromsun.html)

theory seems to work for Mercury and Mars, but not so well for Earth and not at all for Venus. The reason is the greenhouse effect, to which we now turn.

The Greenhouse Effect

Earth's atmosphere contains gases that absorb longwave radiation, but not shortwave radiation. I defer the details until chapter 3, where we analyze the atmosphere as the medium of weather. Here the bare physical fact is enough: the atmosphere is transparent to most shortwave radiation and opaque to most longwave radiation. This state of affairs leads to the greenhouse effect. Mercury has no atmosphere, and Mars has only a tenuous atmosphere; this is the reason why their actual surface temperatures are close to the blackbody value, as we saw in figure 1.3. They can have no greenhouse effect.

Planets with atmospheres that absorb longwave radiation (Earth and, especially, Venus) receive shortwave radiation from the sun. Some of this radiation is reflected back into space, but most passes through the atmosphere and is absorbed at the surface of the planet. The planet is nearly a blackbody in equilibrium, and so it reradiates the same amount of power that it absorbs, but as longwave radiation. This long-wave radiation emanating from the surface of the planet is absorbed by the planet's atmosphere—partially in the case of Earth and almost wholly in the case of Venus. Consequently, the planetary atmosphere heats up (because it is absorbing more radiation than it is emitting). This is the greenhouse effect.

We may regard Earth and its atmosphere as a single entity, a black-body that emits as much radiation as it absorbs. The temperature of the "Earth-plus-atmosphere" blackbody is higher than that of the "Earth" blackbody because of atmospheric absorption of longwave radiation. More details are given in chapter 3.

Heat Transfer

Here is the story so far: our sun emits EM radiation that heats up Earth according to well-established thermodynamic laws. The mean surface temperature of Earth is a little higher than these laws alone predict because the greenhouse effect, an atmospheric phenomenon, kicks in.

So we residents of the planet, on its surface and under its atmosphere, are living in a zone of dynamic equilibrium. Vast amounts of heat are pouring in as shortwave radiation; equally vast amounts are pouring out to space as longwave radiation. We are in the middle, at +15°C (59°F) on average. We will see in chapter 2 how this situation leads to weather; here we prime the pump by looking into the means by which heat can be transferred from one body to another—how it can be shifted from place to place over the surface of Earth.

The three mechanisms of heat transfer, which you can read about in any elementary physics text, are

- Advection
- Conduction
- Radiation

Advection is the physical movement of a hot body. No, this isn't a boyfriend or girlfriend walking across the room, but the movement of heat due to the movement of the object that holds it. Hot water flowing through a pipe in a house transfers heat by advection—the heat is in the water, which moves. Okay, I guess a boyfriend or girlfriend walking across a room is an example of advection, if he or she has a body temperature that is greater than that of the room. Convection is a type of advection; it is the vertical transfer of heat from one place to another as a result of fluid motion. We will see many examples throughout this book of heat at the surface of Earth being transferred to higher altitudes by rising air. The surface heats the air next to it (conduction); the heated air is less dense than the air around it and so it rises, taking its heat with it (convection).[7]

Conduction (also known as *diffusion*) is the transfer of heat via physical contact. A hot rock placed in cold water will cause the water to heat up. A poker that is red hot at one end will warm up at the other end, due to conduction of heat along the length of the poker.

We have already encountered radiation. It *is* heat. It is EM power that moves at the speed of light (in space, actually a little slower through the air). That red-hot poker may slowly heat your hand by conduction through the poker body, but it heats your face much faster by radiation. Red light (and unseen infrared radiation) emitted by the hot end of the poker travels through the air and is absorbed by your face.

A domestic heating system involves three of these heating mechanisms. The water heater heats water by conduction; the heated water is physically moved to the radiators (advection), which then heat a house by—no surprise—radiation.

So much for the mechanisms of heat transfer; we will meet examples of all these mechanisms in the pages that follow. What about the body that receives the heat? How much does it heat up? We know that different bodies heat up at different rates. Thus—and here is an example that matters for weather physics—on a hot day at the beach, the sand heats up much more than the sea, even though both receive the same rays from the sun. The reason is that sand and water have different *heat capacities*. Heat capacity is a physical property of substances; it is the ability of the substance to absorb heat. The heat capacities of glass, steel, aluminum, and air are all different; some relevant examples are shown in figure 1.4. It turns out that water has one of the highest heat capacities of any common substance (among which, only that of ammonia is higher), meaning that it takes a large amount of heat to

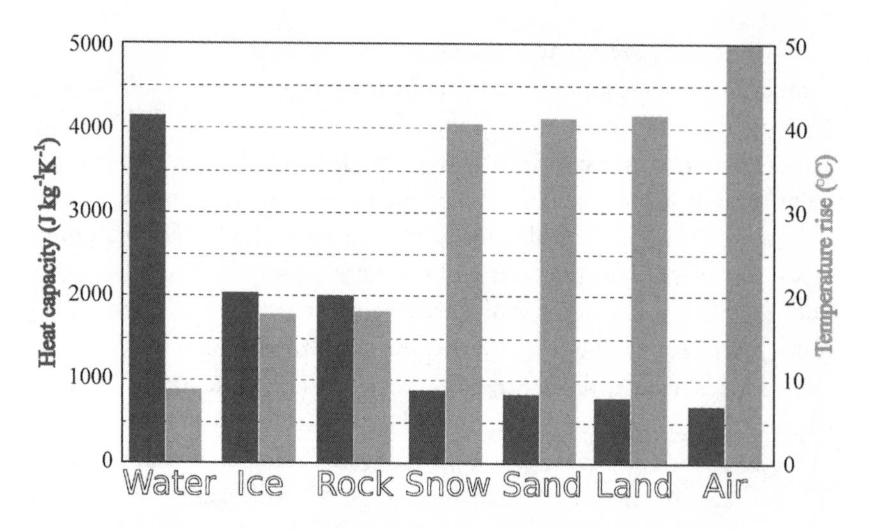

Figure 1.4 Heat capacity of seven natural substances (*left scale*). The value for rock is an average of many different types; the values for sand and air assume that they are dry; land includes vegetation cover. Heat capacity can be defined as the ratio of heat input to consequent temperature rise. If we take 100 kilograms (220 pounds) of each substance, and heat it up with 1 kilowatt of power for an hour, then it will rise in temperature by the amount shown (*right scale*). Note how water can soak up a large amount of heat, compared with everything else.

make its temperature rise much. One consequence is that the oceans of the world can soak up a huge amount of heat. This fact has enormous consequences for both weather and climate.

Other factors influence the amount of heat that is absorbed by a body. Consider a body of water. The reflection of EM radiation off the surface of the body depends on the angle of incidence—more is reflected and less is absorbed for a small, glancing angle than for a large one. So, for example, we can expect that the water will be heated more at midday than later or earlier in the day. Also, the transparency of water influences absorption—the depth to which radiation will penetrate. Another factor to consider is that vegetation greatly influences the amount of EM radiation that is absorbed. After all, leaves are meant to catch sunlight and turn it into energy for the plant.

In chapter 2, we will see the mechanisms by which the heat energy of Earth's surface leads to active weather. One such mechanism is as clear as night and day: Earth spins on its axis, and so the heat delivered to a given point on the surface is not constant. Heat differences arise, and these lead to fluid movement. Air flow and ocean currents are the beginnings of weather; indeed, they *are* weather.

I need to introduce you to one other aspect of heat transfer before we can leave this subject. Water at boiling point (100°C [212°F] at standard atmospheric pressure) can turn into steam—change phase from liquid to gas. But this change of phase takes energy (heat) and a large amount of it. This energy is called, logically enough, the *heat of vaporization* of water. To change 1 kilogram (2.2 pounds) of water at 100°C into steam at the same temperature requires 2.257 megajoules of energy. That's the energy from a 2-kilowatt electric heater acting for 19 minutes. When the steam condenses to re-form water, it *releases* this amount of energy; which is why cyclones can pick up energy from warm oceans, as we will see. Say the steam cools from 100°C to air temperature—we call the water content of this cooled gas *water vapor*—then the vaporization energy is still locked up inside the molecules, even though the vapor feels cool to the skin. This type of energy is termed *latent heat.*

***

The sun emits huge amounts of EM energy from its surface, originating with thermonuclear reactions in the interior. The orbiting planets are bathed in this energy and achieve an equilibrium temperature that can be estimated thermodynamically. The actual surface temperature of Earth (and especially of Venus) is higher than the blackbody temperature because of the greenhouse effect. Heat is transferred around Earth's surface by several mechanisms; much is held in the oceans because of the very large heat capacity of water. The heat held within our oceans, land, and atmosphere drive our weather.

2

Under the Heavens and the Seas

To be interested in the changing seasons is a happier state of mind than to be hopelessly in love with spring.

George Santayana

There are fast and slow changes in the detailed position of Earth under the heavens; these physical changes profoundly influence weather and climate, respectively. In addition, geology influences climate, and geological changes lead to climate changes. In this chapter, which is introductory in the sense that it anticipates many topics discussed later in the book, we see how the physical geography of Earth influences the delivery of solar heat to (and around) the surface.

Surface Irradiance and Surface Features

We saw in chapter 1 that the level of electromagnetic (EM) solar power arriving at the top of our atmosphere amounted to about 1.7×10^{17} watts. This huge quantity can be comprehended better by converting it into a power density: 1,365 W m^{-2}—that is, 1.365 kilowatts for every square meter of surface. Note from figure 2.1 that the relevant surface here is that of a circle with the radius of Earth, located at the top of our atmosphere. From the figure, it is clear that the average

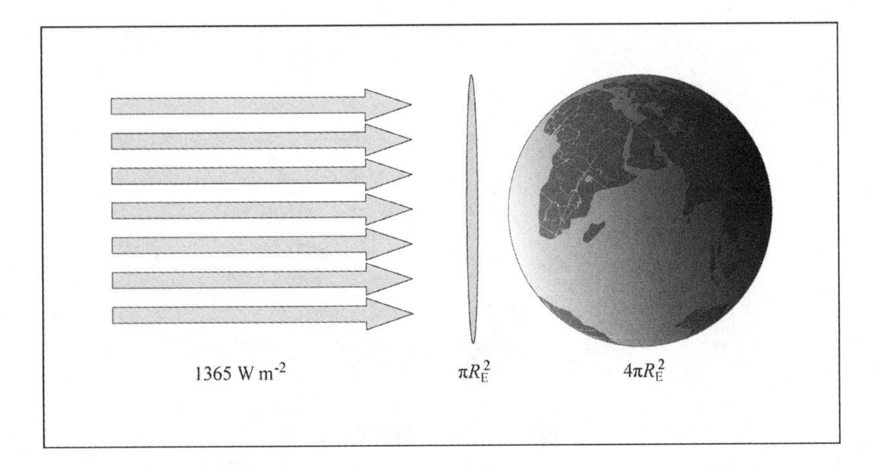

Figure 2.1 Solar radiation that reaches Earth passes through an imaginary circle with Earth's radius, R_E (shown edge on) at the top of the atmosphere. This radiation covers the surface of the planet over a 24-hour period, so the mean power density at the surface is a quarter of the mean power density at the imaginary circle. (The surface area of a sphere is four times the area of a circle of the same radius.)

power density on the surface of the planet is less than this amount by a factor of four. Thus the power density we will assume, when we turn to detailed investigations of the incoming shortwave power density, begins at a mean value of 341 W m^{-2}.[1]

This number is an annual average for the whole planet surface. The solar radiation received at any given point on the surface varies with time of day (it is zero at night, of course). The daily average varies with latitude and with season. Seasons arise because of the tilt of Earth's rotation axis (figure 2.2). This tilt is known precisely (it is 23.4°), and so we can calculate quite accurately the solar power density (the *irradiance*) that would reach any point on the surface if we could ignore the effects of the atmosphere. The influence of our atmosphere is discussed in detail in chapter 3; here let us simply note that, due to reflections and to absorption (figure 2.3), the solar power reaching the surface is attenuated by about 30% compared with the power at the top of the atmosphere.

The picture that is forming from the information just provided is one of differential heating over the surface. The key component here

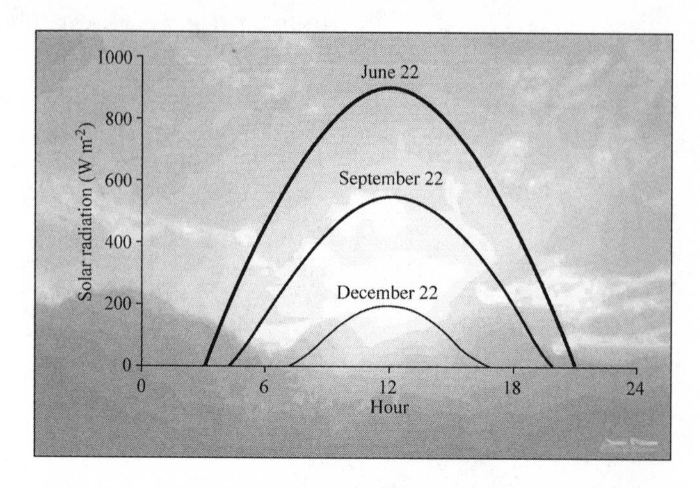

Figure 2.2 Expected power density received on the surface of Earth at latitude 52°N in southern England during the course of a day, assuming no cloud cover (a big assumption, given English weather). Hour 0 corresponds to midnight, and hour 12 to midday. Note that the power densities vary considerably with season as well as with hour. (Adapted from P. Burgess, "Variation in Light Intensity at Different Latitudes and Seasons, Effects of Cloud Cover, and the Amounts of Direct and Diffused Light" [paper presented at Continuous Cover Forestry Group Scientific Meeting, Westonbirt Arboretum, September 29, 2009], fig. 5, http://www.ccfg.org.uk/conferences/downloads/P_Burgess.pdf)

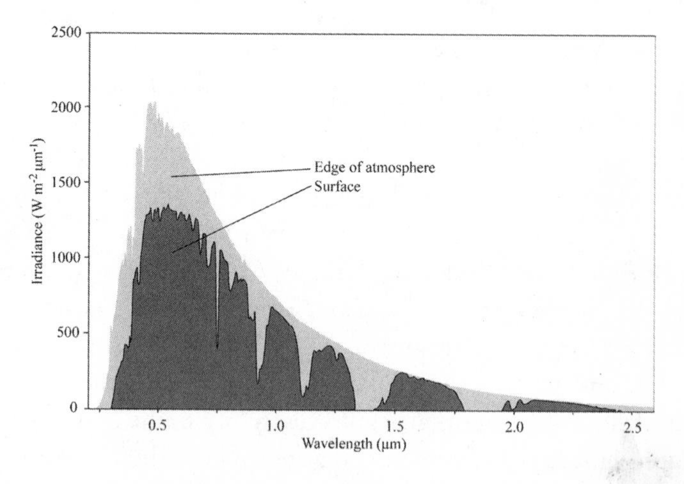

Figure 2.3 Solar radiation traveling down through the atmosphere is attenuated by absorption as well as by reflection. Note how the absorption depends on wavelength (later we will see that water vapor is the main atmospheric constituent responsible). This graph is for shortwave radiation; a few wavelengths are strongly attenuated, but much of the spectrum reaches the surface. (Adapted from a graph by Robert A. Rohde)

is Earth's 24-hour rotation: imagine a chicken on a spit getting cooked on one side and slowly rotating. But the differences are more widespread and less uniform than this image suggests. The surface of our planet is 70.9% water. The remaining 29.1% is land composed of thermally inhomogeneous material, varying from dark rock to pale sand. All these surface components reflect EM radiation differently, and so the power they absorb varies (box 2.1). Thus the world is heated by shortwave solar power, but the amount of heating depends on the time of day, latitude, and physical details of the surface being heated.

So what is the consequence of the fact that Earth's surface is not heated uniformly? Weather. This simple statement is the bottom line of this chapter. Weather is what happens when the heating of Earth is not constant—not uniform over the surface (how could it be uniform?) and not constant in time. Here is an obvious, simple, and very important example that we will unpack in more detail later on: the equatorial

Box 2.1
Whiteness

The reflectance of a substance is the fraction (or percentage) of incident radiation that is reflected off its surface. The remainder is absorbed (or transmitted through the substance in the case of transparent substances). Reflectance is historically known as *albedo,* from the Latin for "whiteness." The albedo of a given substance depends on many factors, such as its chemical composition and structure, the angle of incidence of the radiation, the radiation frequency, and the surface roughness. The albedos of several components of Earth's surface, and of Earth as a whole, are shown in box figure 2.1.

The importance of albedo for climate change can be seen by considering polar ice, which has a higher than average albedo. During an ice age, polar ice forms, which increases the albedo of the planet. Increased albedo means that Earth absorbs less heat and so its temperature reduces, which causes more ice to form, which increases planetary albedo further, and so on. This is an example of *positive feedback.* The reverse process can also occur; that is, higher temperatures melt ice, which reduces albedo, which increases temperatures further—also positive feedback. Feedback mechanisms abound in climatology, as we will see in chapter 4.

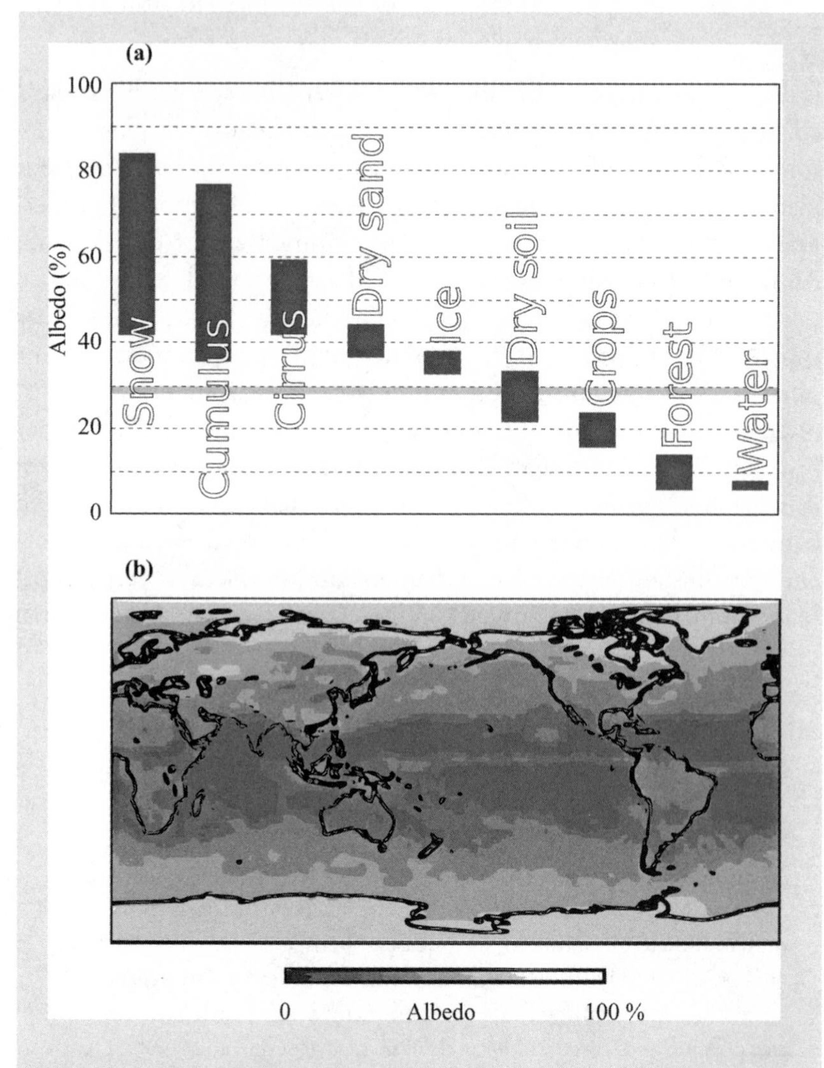

Box Figure 2.1 Reflectance/albedo. (*a*) Albedo for several naturally occurring substances. Albedo is higher (reflectance more complete) for fresh snow than for older snow. Albedo is high for clouds and varies between and within cloud types. Crop land and forests have low albedos because the function of leaves is to absorb light (and, through photosynthesis, convert it into energy that the plant can use). The albedo range shown for water assumes high angle of incidence—at practically a right angle. At low (grazing) angles, water has a very high albedo. The average albedo of Earth is shown by the horizontal gray line. (*b*) Albedo of Earth. Note that more radiation is absorbed in equatorial regions than in temperate or polar regions. ([*b*] Adapted from an image in NASA, "Global Albedo," Visible Earth, www.visibleearth.nasa.gov/view.php?id=60636)

surfaces of Earth get heated more than temperate surfaces, and so the atmosphere and oceans near the equator are warmer than the atmosphere and oceans farther north or south (figure 2.4). We will see that this simple fact of orbital geometry leads to atmospheric circulation patterns and trade winds, to the doldrums and ocean currents, to equatorial rainforests and deserts on continental landmasses immediately north and south of the equator.

The surface of our planet is nonuniform in other ways, of course, principally in the uneven distribution of landmasses. Sixty-eight percent of land is in the Northern Hemisphere and only 32% is south of the equator. If we divide Earth into longitudinal hemispheres, with the predominantly Eastern Hemisphere covering longitudes 20°W to 160°E, then the Eastern Hemisphere also contains 68% of the landmass, coincidentally. Thus viewing the disk of Earth from the sun, we see throughout any 24-hour period a varying surface, sometimes mostly land and sometimes mostly ocean. The albedos of these surfaces differ markedly, and so the amount of solar power absorbed by the planet varies markedly throughout the day. It similarly varies from season to season. Differential heating over the course of a day, you see, and so daily weather.

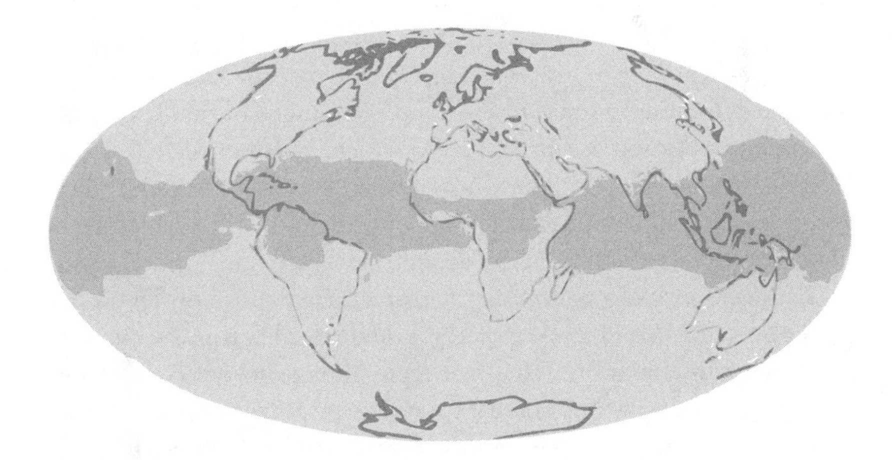

Figure 2.4 Hotspots, or areas where the incoming shortwave radiation power exceeds the outgoing longwave radiation power by at least 100 W m^{-2} (*dark shading*). Thus equatorial regions receive more power than the world average. Further data show that polar regions receive less than average. (From NASA, from data gathered in September 2008)

The distribution of continents influences wind speeds. Thus the Southern Hemisphere contains the Roaring Forties and the Furious Fifties, latitudes infamous in the days of sailing ships for damagingly high wind speeds. These high wind speeds are due in large part to the paucity of land at these southern latitudes. (Resistance to wind flow is less over water than over land.)

The oceans are deeper than the land is high. That is, the average elevation of land across the world is 840 meters (2,760 feet), whereas the average depth of the oceans is 3,700 meters (12,100 feet). Hence we live on mountaintops, even if we live at sea level. There is a *lot* of water on the surface of our planet—some 1.35 billion cubic kilometers (320 million cubic miles). We have earlier noted the large heat capacity of water and the large surface area of the oceans. Putting all these facts together, it is immediately apparent that the oceans receive a large amount of solar power and can hold a huge amount of heat. It is not difficult to show that if the oceans absorbed all the solar power that fell on them, and did not reemit this power as longwave radiation, then it would take the sun two years to heat up the oceans by 1°C (33.8°F). Of course, this calculation is unphysical because thermodynamics demands that the absorbed power is reradiated almost entirely, but it serves to show that the oceans constitute an enormous heat reservoir. To anticipate the discussion of global warming in chapter 4, consider what would happen if the atmosphere were to heat up suddenly (geologically sudden, say over a couple of hundred years). We can expect that much of this heat would be soaked up by the oceans and so we would not immediately see all of the consequences of the sudden warming. For better or worse, these consequences will be spread out in time owing to the thermal inertia of climate systems, arising from the oceanic heat reservoir.

The depth of many oceans means that their resistance to flow is low (compared with that of shallow waters), and this facilitates ocean currents. Ocean currents are a very important topic in meteorology, perhaps surprisingly at first sight but not after a little consideration of the physics. They merit a separate section in this chapter, and so I will defer further discussion. The large surface area of our oceans means that water can readily evaporate into the atmosphere—another key aspect of Earth's weather. Water vapor is a greenhouse gas, the stuff of clouds (which greatly influence Earth's albedo, as suggested in figure 2.5),[2] of rain and snow, and of heat transport to the higher reaches of the troposphere, as we will see.

Figure 2.5 Cloud cover over Earth, March 11, 2005. Note how the clouds increase the planet's albedo. (From NASA's *Terra* satellite)

Ninety percent of the mass of our planet is made up of four elements: iron, oxygen, silicon, and magnesium. Almost all the iron is in the interior, whereas almost all the oxygen is near the surface. Life has harnessed oxygen in a big way and has changed our planet as a consequence. Thus life is responsible for freeing oxygen from Earth's surface, so that it is now the second most common component of our atmosphere. This was not always so. Trees are, from the perspective of a meteorologist, very efficient machines for turning groundwater into atmospheric water vapor. From the perspective of a geologist or climatologist, they are important machines for fixing carbon (box 2.2). The evolution of trees in particular, and plant life in general (plus, recently, one type of animal), has changed the physical surface of our planet, its climate, and its weather.

Geological changes in landmass distributions have also led to significant changes in the climate of our planet over the last several hundred million years. Thus, for example, in past eons the landmass has twice been concentrated almost entirely in one hemisphere (750 million years ago and 240 million years ago). The distribution of continents greatly influences global oceanic currents, which are a crucial component of our weather and climate. The fraction of Earth's surface that has been occupied by water has varied over the eons between 70% and 87%;

Box 2.2

The Carbon Cycle(s)

There are two timescales over which carbon is cycled through the ground, the oceans, and the atmosphere. The slow cycle takes on the order of 100 million years: carbon dioxide is absorbed by atmospheric water vapor to form a mild acid, which falls as rain and dissolves rocks, which flow into the seas and form sediments, which are compacted into rocks by geological processes, which are spewed out into the atmosphere by volcanoes. This summary is brutally abbreviated because the slow carbon cycle is only of peripheral interest to us. It locks up the vast majority of cycled carbon in the lithosphere (rock layers of Earth's crust), though the amount of carbon cycled per year is much less than that of the fast cycle. The slow carbon cycle has been short-circuited by human activity, operating on a much faster scale, for example, by burning fossil fuels. A branch of the slow cycle acts on a less-slow timescale and is more relevant. There are regions of the oceans where deep carbon-rich waters well up to the surface and release carbon dioxide to the air. The amount of carbon released in this way is naturally balanced, neglecting human activity, by carbon that enters the seas via runoff from weathered rocks.

The fast cycle takes place on the scale of life—say, a year. Between 1 billion and 100 billion tons of carbon per year (the rate is highly variable) are fixed in the bodies of plants and phytoplankton* by photosynthesis—the conversion of atmospheric carbon dioxide into glucose, fueled by the energy of sunlight. The carbon then disperses to other life-forms as herbivores eat the plants, carnivores eat the herbivores, and so on. When a tree or an animal dies, it decays as a result of the actions of bacteria and fungi that cause most of the captured carbon to be released back to the atmosphere. The fast cycle can be seen as an annual variation in atmospheric carbon dioxide levels: carbon is fixed in spring and summer as trees leaf up, so the carbon dioxide level drops. Then in fall, the leaves drop and the level of atmospheric carbon dioxide increases again.

Atmospheric carbon takes the form of carbon dioxide and methane—both greenhouse gases. Although carbon dioxide contributes less to the greenhouse effect than does water vapor, it sets the temperature by regulating the atmospheric water vapor content, as follows. An increase/decrease in atmospheric carbon dioxide causes an increase/decrease in air temperatures, which leads to a higher/lower water vapor content in the atmosphere. Thus human-made changes in the natural level of atmospheric carbon dioxide have consequences for global warming. The effects of a sudden rise in atmospheric carbon are largely but not entirely mitigated by a consequential increase in absorption of carbon by the oceans. In the words of one study:

"[T]here is no natural 'savior' waiting to assimilate all the anthropogenically produced CO_2 in the coming century."†

*Phytoplankton are microscopic organisms that live at the surface of lakes and oceans, which generate the energy they need for life by photosynthesis.

†P. Falkowski et al., "The Global Carbon Cycle: A Test of Our Knowledge of Earth as a System," *Science* 290 (2000): 291–296. There are useful nontechnical accounts of the slow and fast carbon cycles in National Aeronautics and Space Administration, "Carbon Cycle," NASA Science/Earth, http://science.nasa.gov/earth-science/oceanography/ocean-earth-system/ocean-carbon-cycle/; and National Oceanic and Atmospheric Administration, "Carbon Cycle Science," Earth System Research Laboratory, http://www.esrl.noaa.gov/research/themes/carbon/. See also J. T. Houghton et al., eds., *Climate Change 2001: The Scientific Basis; Contribution of Working Group 1 to the Third Assessment Report of the Intergovernmental Panel on Climate Change* (Cambridge: Cambridge University Press, 2001).

the lower end occurs during ice ages, such as the one we are in,[3] which lock up water in polar ice sheets. Polar ice sheets also greatly influence global oceanic currents.[4]

Long-Term Orbital Effects

The simple geometry of Newtonian orbits—fixed ellipses—applies in reality only for a solar system with one spherical planet and no moons. (Even in this case, the sun and the planet would have to be very small, or not rotating.) Reality is more complex, and so the orbital characteristics of Earth are more complicated than for the simple case that we all learned about in high school. The real Earth and sun rotate and so are not spherical, but bulge in the middle. We have a large moon. We have sister planets—fellow travelers around our star. All these facts create a major headache for astronomers who want to know the detailed shape of our orbit: the calculations are far, far more difficult than those that Sir Isaac Newton taught us, even though the predictions they make are only very slightly different from his.[5]

We will need to understand only the nature of these complications, not their derivation. The important point to note is that all the orbital parameters that apply to Earth are variable. The shape of the orbit changes over time; its orientation is changing; the length of a day is changing; the north–south axis, about which Earth spins, is not fixed.

All these changes are small and slow by everyday standards, but over millions of years they add up significantly and they change our climate. Climatologists now understand much, but not all, about how these astronomical perturbations have influenced past climates. The perturbations may be small, but, because they act for such a long period of time, they can have very significant effects on global climate. As one prominent example, these small complications are responsible for Earth's ice ages.

The long-term orbital phenomena that influence our climate are

- Precession of the rotation axis (precession of the equinoxes)
- Orbital plane wobble (obliquity of the ecliptic)
- Perihelion (or apsidal) precession
- Orbital eccentricity oscillation

A little explanation is needed because astronomy has an old and complicated vocabulary. Axial tilt is clearly responsible for the seasons. Thus when the Northern Hemisphere is facing the sun, it experiences summer (figure 2.6*a* [*left*]); at the opposite point on the orbit, it experiences winter (*right*). Orbital eccentricity alters this symmetrical arrangement a little. Currently, the perihelion (the closest approach of Earth to the sun) occurs on January 3 and the farthest point on July 4. It may seem strange to residents of the Northern Hemisphere that the planet is closest to the sun during winter, but the difference is very small (1.6%). Nevertheless, it does mean that the northern summer is 4 days and 16 hours longer than that of the southern summer.[6] Variation in eccentricity leads to variation in solar power levels and in lengths of the seasons. Precession of the perihelion also causes small changes in the lengths of the season. The combination of perihelion precession and precession of the equinoxes means that the perihelion returns to the same date every 21,600 years.

The possibility that these small and slow orbital perturbations influence climate by changing insolation (received solar power) and by varying season lengths was proposed a century ago by a Serbian astronomer, Milutin Milankovitch, who began his work on this subject in 1915 while languishing in a World War 1 prisoner-of-war camp.[7] His work was published but was not accepted by the scientific community as a true description of nature until after his death; Milankovitch was

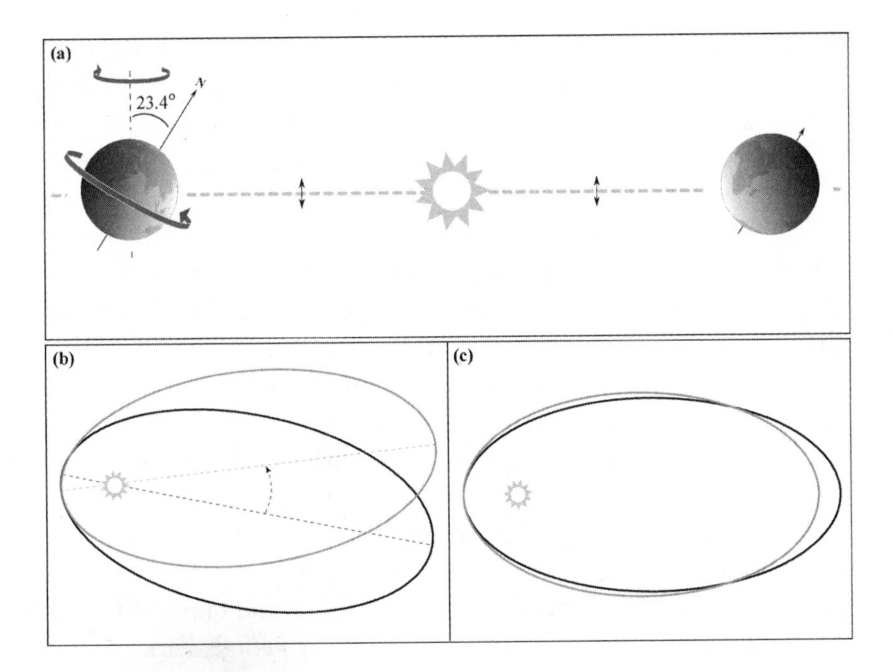

Figure 2.6 Orbital wobbles. (*a*) We have seasons because the north–south rotation axis (N) is tilted, currently by 23.4°. The direction of rotation itself slowly moves around a circle, as shown, taking 25,770 years to complete each loop. The orbital plane (*dashed line*) changes orientation, which has the effect of changing the tilt angle; it oscillates between 22.0° and 24.5°. The oscillation period is 41,000 years. (*b*) The long axis of Earth's orbit revolves about the sun and takes 134,000 years to make one revolution. (*c*) The eccentricity of the orbit varies between 0.00 and 0.06, with a period of 90,000 to 100,000 years. Here the eccentricity of Earth's orbit has been exaggerated to about 0.87 (the actual eccentricity is so small that the orbit looks like a circle).

ahead of his time, and, in any case, there were insufficient data to back up his assertions. In the 1970s, researchers extracted from deep-sea cores global temperature histories going back millions of years. These histories displayed periodicities of 23,000 years, 43,000 years, and 100,000 years, in close agreement with the periodicities of the long-term orbital cycles now known as Milankovitch cycles. Further, the periodicities line up with known events such as the occurrence of ice ages. Milankovitch's theory is now widely accepted, though our understanding of the mechanism is incomplete and the data are not entirely in agreement with the theory. Thus, for example, the 100,000-year period of ice ages seems to be a very big animal in the climate history zoo and yet is a relatively small creature in the menagerie of Milankovitch cycles. Physicists and climatologists are still working on the theory.[8]

In addition to these orbital forcing terms, as the Milankovitch contributions to climate change are collectively termed, there are other long-term astronomical phenomena that are influential. Because we have a moon and oceans, we have tides. The mass movement of our oceans, responding to the combined gravitational pull of the moon and the sun, dissipates heat. The heat generated by tidal movements varies with location because of the landmass distribution and can be as high as 30 kilowatts per square kilometer in the North Atlantic. A consequence of this loss of energy is that the period of rotation of our planet is increasing: the length of a day is increasing by 2.3 milliseconds per century. Although undetectable within the hustle and bustle of human existence, over eons it adds up: in much earlier periods Earth rotated about its axis once every 22 hours.[9]

Water Cycle

Another important aspect of physical geography that we need to appreciate is the water cycle. Water first reached the surface of Earth 3.8 billion years ago, due to volcanic activity, and since then the amount of water on, above, or just beneath the surface has remained fairly constant. This water takes several forms. For every liter of freshwater, there are 40 liters of saltwater. More than two-thirds of the freshwater is held in glaciers and ice caps, 30% is groundwater, and the rest—only 1.2%—is surface water. Of the surface water, 69% takes the form of ground ice and permafrost, 21% is pooled in lakes, and the remaining 10% is in the soil, in rivers or marshes, or in the atmosphere.

Thus the amount of atmospheric water is a minuscule fraction of the whole and yet it is the crucial component of global weather. There is a cycle of water molecules between the various forms (figure 2.7). This water cycle is driven by the heat of the sun.

The mechanism is quite simple. Solar power heats up surface water, mostly in the oceans, which evaporates into the atmosphere. (A small

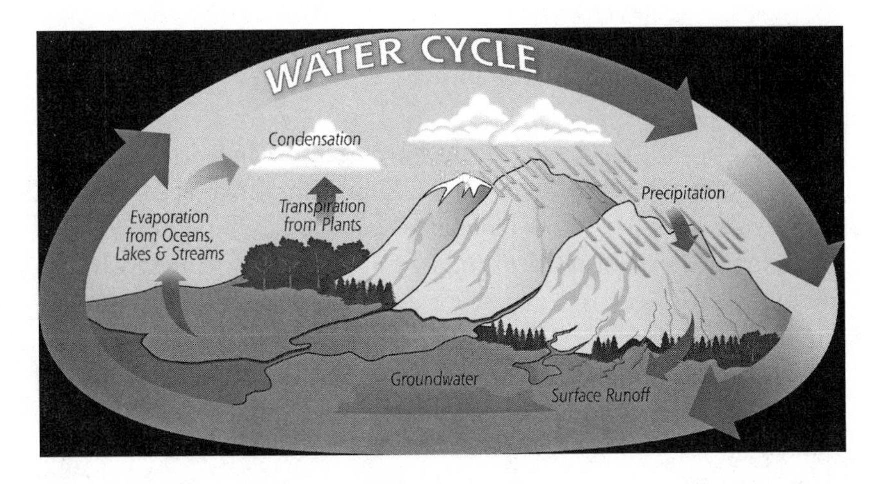

Figure 2.7 Simplified schematic of Earth's water cycle. (From NASA, "The Water Cycle," Precipitation Education, https://pmm.nasa.gov/education/water-cycle)

but significant extra source of atmospheric water vapor is provided by the transpiration of groundwater through the leaves of plants and trees.) The air containing this water vapor rises and cools. At a certain altitude, the temperature is cool enough to cause the water vapor to condense, forming clouds. Winds may blow the clouds over land. Further cooling causes the clouds to shed precipitation either as water or snow. Rain soaks into the ground and adds to the water table, or it pools into lakes or runs off the surface in rivers back to the seas and oceans. The quantity of water moving through the water cycle is mind boggling: approximately 500,000 cubic kilometers (121,000 cubic miles) of water evaporates and precipitates each year. Expressed in a scale more easily comprehended, this amount is equivalent to an average evaporation/rainfall rate of about 110 grams (say, 4 ounces) of water per hour for every square meter of surface.

The time spent in each part of the cycle is very variable and depends on the phase of the water. Solid water in the form of Antarctic ice is locked up for an average duration of 20,000 years. Liquid water in the oceans remains there for an average of 3,200 years but for only 100 years in lakes and a couple of months in rivers. Gaseous water, the vapor in our atmosphere, likely will remain as such for only nine days before moving on to the next part of the water cycle.[10]

Oceanic Circulations: Heat Pumps

Each section of this chapter could have been expanded into a whole book; this section could have been a library shelf. We need to touch on only those aspects of oceanography that have an impact on weather and climate, but that restriction does not cut down the field as much as you might think. We cannot, for example, concentrate only on the surfaces of our seas and oceans because what goes on beneath has enormous implications for us. The ocean deep is coupled with the ocean surface: a significant amount of water is exchanged between the two, though only at certain locations. (Why? Read on.) Of necessity, I will skim the surface of oceanography, if not of the oceans, and concentrate on physical principles and processes pertaining to weather and climate.

Forces

Oceanic currents operate on all length scales, from gyres that cover thousands of kilometers to eddies that are a few centimeters across. They are all ultimately driven by solar heat, but with different proximate causes: winds, pressure gradients, salinity gradients. The shape and direction of very large-scale circulations are influenced by the rotation of Earth. How so, you reasonably ask. Rotation influences water (and air) circulation by contributing a force, the mysterious *Coriolis force*. In chapter 8, I fully unwrap and explain this oddity, which is a significant player in the circulation game but only for movements of planetary scale (it is negligible on smaller scales). Here we need to know that this force acts only on moving masses, that the magnitude of the force is proportional to the speed of the mass, that it increases in magnitude from zero at the equator to a maximum value at the poles, and that it acts at right angles to the direction of motion. In the Northern Hemisphere, it pushes a moving mass to the right, and to the left in the Southern Hemisphere. Very odd.[11]

Other relevant forces are more familiar. When an ocean current piles up against a landmass, the water level rises: the surface exhibits a measurable slope. Thus the pressure at a given depth is greater nearer the landmass than farther away, and so there exists a horizontal force, due to this pressure difference, that pushes the water away from the land.

Seawater density varies with temperature and with salinity (saltier and cooler waters are more dense than fresher or warmer waters). The force of gravity thus causes cool or salty water to sink if the surrounding water is warmer or less salty. Gravity, this time that of the moon and sun, also acts to produce tides. Tides are not part of the ocean circulations that greatly influence weather—though weather can greatly influence tides—and so we do not look into the subject of tides any further.

Winds blowing across the surface of a sea or an ocean can push the surface water due to frictional coupling between the two fluids. The frictional force (known as *wind stress*) that acts between air and water is quite complicated because the physics of fluids is quite complicated. The force acting on the surface water increases as the square of wind speed and increases with the turbulence of the water surface. It increases with the area of water that is influenced by the wind—particularly by the *reach*, which is the linear distance along the wind direction over which the wind blows. The depth of water that is affected by atmospheric winds varies with seawater conditions, as we will see. Empirically we see that ocean current speed is typically about 3% of the wind speed that drives it.

If the horizontal forces that act on water—say, wind stress at the surface—diverge, then this can lead to vertical movement of the water at the center of the divergence (figure 2.8*a*). Divergent forces push water away from the center and so deeper water wells up to fill the gap. (Similarly water in the center will sink under the influence of convergent forces.) In figure 2.8*b*, we see how wind and water interact at the equator when large-scale divergent forces combine with the Coriolis force to produce an interwoven pattern of atmospheric and oceanic circulations. This example provides an indication of the interactions between wind and water that make global meteorology complex, interesting, and challenging.

LAYERS

The ocean is layered vertically like a cake (and like the atmosphere, as we will see in chapter 3). The division into layers is based on salinity and temperature gradients, and so it is variable. Salinity at the surface changes, for example, when it rains or when hot weather induces more than the usual amount of evaporation. Salinity also varies spatially; it is

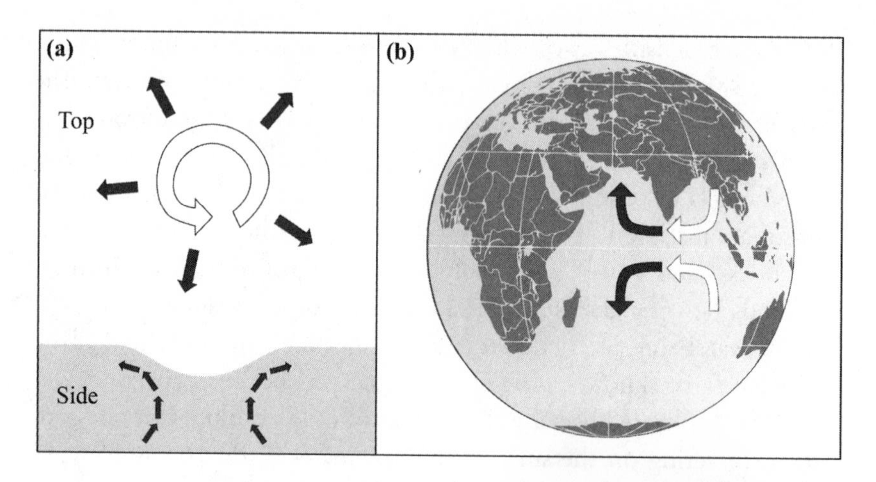

Figure 2.8 Divergent wind stress. (*a*) Cyclones arise over open water. Here the circulating winds (*white arrow*) push surface water (*dark arrows*) outward, causing an upwelling from below the surface. Water may be moved vertically perhaps 100 meters (330 feet), and horizontally perhaps 100 kilometers (60 miles). The effect of the cyclone on the water underneath it is to form a current loop in the shape of a flattened donut. (*b*) On a global scale, equatorial heat causes air to rise, which pulls in air from farther north and south to the equator. In the Northern Hemisphere, the Coriolis force turns the moving air to the right—westward—while in the Southern Hemisphere, it is turned to the left—also westward (*white arrows*). Hence our westward equatorial winds. These winds exert stress on the surface water beneath them, so these waters are pushed westward but also diverge, as shown (*dark arrows*), because of the Coriolis force. The diverging waters lead to upwelling of deeper water to the surface, near the equator.

low near the mouths of major rivers and at the poles (where evaporation is reduced). Temperature profiles vary with latitude, water depth, and local weather conditions. In general, we can say that there is a shallow surface layer of a few hundred meters sitting on top of a *thermocline*, which is a layer where temperature rapidly falls with depth, between the well-mixed surface layer (where temperature is constant) and a colder body of water beneath. Sometimes, if the surface layer is exceptionally warm or fresh (that is, "unsalty," to coin a word), then the thermocline is very thin and acts like a lubricant, in that there is little frictional coupling between surface water and deeper water. Oceanographers call this a "slippery sea" because wind stress causes water movement only at the surface. More usually, turbulence mixes the surface waters with deeper waters, and so wind stress at the surface induces movement that carries down, albeit attenuated, into the cooler depths.

The vertical layering of oceans is much more marked than any horizontal variation in water characteristics, which usually depend on only seabed depths and composition. The reason for vertical but not horizontal layering is simply gravity: denser fluids sink, and so a water column becomes separated based on temperature and salinity differences, which change water density. Vertical movement is possible but is more constrained than horizontal movement, so there is less vertical mixing.

We have seen that over long stretches of ocean, the Coriolis force comes into play. Wind stress pushes the surface layer in the same direction, but the Coriolis force then pushes it sideways, resulting in a diagonal movement. Thus the surface layer will move at an angle roughly 45° from the wind direction above it. If turbulence exists between the surface layer and the layer of water immediately beneath it, then this layer will also be induced to move by the wind-driven current above it. Again, the Coriolis force shifts the direction (to the right in the Northern Hemisphere, recall), and so we find that wind-induced currents move in a direction that changes with depth (clockwise in the Northern Hemisphere).

There is a limit to how much the Coriolis force can deflect a moving body of water (or air). North of the equator, a surface current is shifted to the right and so rotates clockwise. But after a quarter turn, the direction of the Coriolis force is then opposing the wind stress that gave rise to the current and so can induce no further deflection. The result is a current that moves at a right angle to the wind. Such a current, where the Coriolis force is balanced by the driving force, is called a *geostrophic* current.

The driving force that gives rise to geostrophic currents does not have to be wind stress; a horizontal pressure gradient will also do the job. The result is that large-scale flows in the oceans are often geostrophic. Geostrophic flow occurs not just in the oceans; you see examples in the atmosphere every time you turn on the Weather Channel. Wind directions that are parallel to isobars on those ubiquitous weather charts (again, I must defer detailed discussion of these, to chapter 10) indicate geostrophic flow.

The most visible currents—those that can be seen and measured by satellites in space—occur in the surface layer of an ocean and cover almost the entire ocean. They employ some 8% of all ocean water. These surface currents are wind driven and are known as *gyres* (whirls, as in "gyrate") and there are five major gyres on Earth (figure 2.9). Each is flanked by a strong and narrow western boundary current, quite different from the weak and broad boundary currents that constitute the eastern flanks. The reason for this difference lies in the wind-circulation patterns that drive the gyres. Note how the sense of rotation of the large circulations is determined by the Coriolis force: clockwise when north of the equator and counterclockwise when in southern latitudes. There can be no gyres crossing the equator because there is no Coriolis force at the equator.[12] Note also from figure 2.9 that the gyres tend to take heat away from equatorial regions and transport it to temperate latitudes. This tendency is particularly important for northwestern Europe because the strong current on the western flank of the North

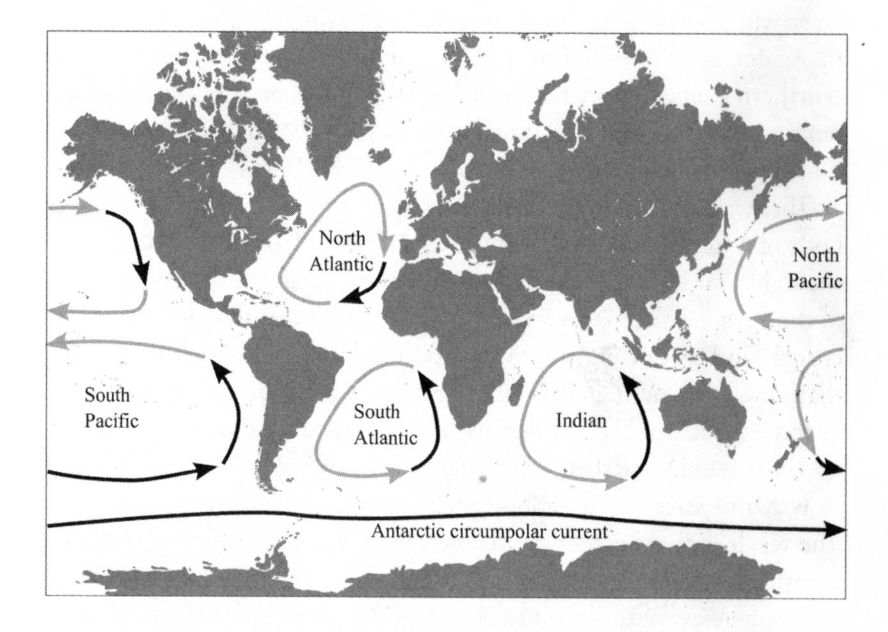

Figure 2.9 The five major gyres cover the five oceans. *Black arrows* indicate cold water and *gray arrows* indicate warmer water. In addition, the Antarctic Circumpolar Current—the world's largest—is shown.

Atlantic gyre is the Gulf Stream, which is likely responsible for keeping this part of Europe significantly warmer than other landmasses at similar latitudes. Originating in the Gulf of Mexico, this current exits the Straits of Florida, continues up the eastern coastline of North America as far as Newfoundland, and then crosses the North Atlantic. It splits into two currents, one heading down toward the coast of West Africa and the other (known as the North Atlantic Drift) warming northwestern Europe. The Gulf Stream as a whole moves at an average speed of 3 knots.

At the center of the North Atlantic gyre is the Sargasso Sea, a broad area of relatively still water that accumulates a large amount of seaweed and other marine debris from the boundary currents.

The Antarctic Circumpolar Current is also shown in figure 2.9. It is the largest in terms of flow. One hundred and twenty-five million cubic meters (4.4 billion cubic feet) of cold water flows past a given line of longitude per second, from the surface to depths of 2 to 4 kilometers (1.2 to 2.5 miles), at an average speed of 2 knots. It is the dominant circulation feature of three oceans and is driven by the strong westerly winds that flow unimpeded by large landmasses at these southern latitudes. This Antarctic Circumpolar Current is thought to have caused the glaciation of Antarctica when it started 30 million years ago, and today acts to redistribute heat between the three southern oceans. It is reckoned to influence significantly the weather in southern Australia, South America, and southern Africa.

Salinity

Deeper layers of the ocean are saltier, in general, because saltwater is denser than freshwater. We have seen that salinity can vary due to freshwater outflow so that local salinity near coastlines reflects inland geography. This is because topography influences rainfall rates and river routes, and together these determine how much freshwater flows off a continent into the oceans, and where it flows off. Thus coastal salinity reflects continental topography.

On a global scale, the salinity of a stretch of seawater depends on its latitude, as is clear from figure 2.10. Equatorial warmth increases evaporation, and so equatorial waters are salty. Polar waters evaporate less, and so are less salty. But polar waters are much cooler, and, in

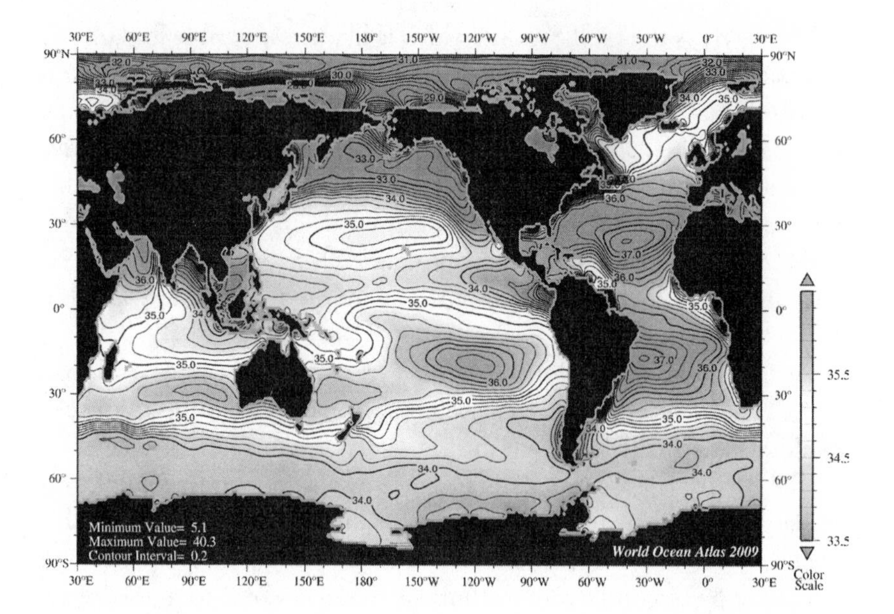

Figure 2.10 Global salinity contours (parts per thousand). Hot regions tend to be salty, rainy regions not so much. (From National Oceanic and Atmospheric Administration)

terms of water density, temperature wins out: polar waters are denser than temperate waters, and so they sink. This contrasts with equatorial waters, which rise, as we saw in figure 2.8*b*. If you see the makings of a vertical current flow here, you are right.

The Mediterranean Sea is shallow (the average depth is only 1,500 meters [4,900 feet]) compared with the oceans, and it is warm. The combination makes the Mediterranean something of a salt pan. This high salinity spills out through the Strait of Gibraltar, and the denser eddies of Mediterranean water (known as *meddies*) tumble down the continental shelf out into the Atlantic Ocean beneath cooler incoming water. (Here is an example where salinity wins out over temperature: the meddies are warmer but denser.) They reach neutral buoyancy at a depth of about 1,000 meters (3,300 feet). Impressively, modern satellite sensors can detect meddies even though they are beneath the surface (figure 2.11). Meddies seem to help maintain the deep ocean currents that are such an important factor in our global climate. We now turn to consider these deep *thermohaline* currents.

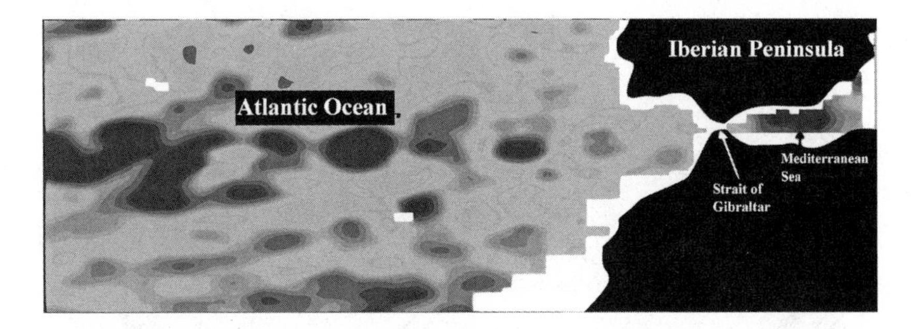

Figure 2.11 Meddies. Salty, clockwise-rotating eddies of warm water (*dark blobs*) spill from the Mediterranean Sea through the Strait of Gibraltar into the Atlantic Ocean. They sink to a depth of about 1,000 meters (3,300 feet). As many as 25 meddies may be detectable at any one time. They act like a salt shaker, maintaining the saltiness of the North Atlantic and so (as explained in the text) maintaining global circulations. (From NASA, https://archive.org/details/meddies_TOP2005anom)

Global Conveyor Belt

The gyres and other surface currents are driven by wind in a direction determined by pressure differences and the Coriolis force. Deep ocean currents, though, are driven by density differences—that is, by temperature and salinity gradients (and the Coriolis force)—hence the technical term for this type of flow: *thermohaline* circulation. More evocative and descriptive is the name given to the interconnected sum of all the world's deep-ocean currents: the *global conveyor belt* (figure 2.12).

The currents of the global conveyor belt are driven as follows. In polar regions, the cold water freezes, leaving saltier water behind. Increased salinity means increased density, and so the water sinks, displacing deeper water and driving the current. Surface water moves in to replace the sinking water. The deep, cold current moves away from the poles, influenced by seabed contours and continental landmasses, forming the currents shown in figure 2.12. These currents are not fast: it may take a bunch of water 1,000 years to travel around the global conveyor belt, corresponding to an average speed of a few centimeters per second. Some branches warm up and become surface currents in equatorial regions. This rising and sinking of water greatly benefits marine life by raising nutrient-rich deep water to the sunlit upper layer.

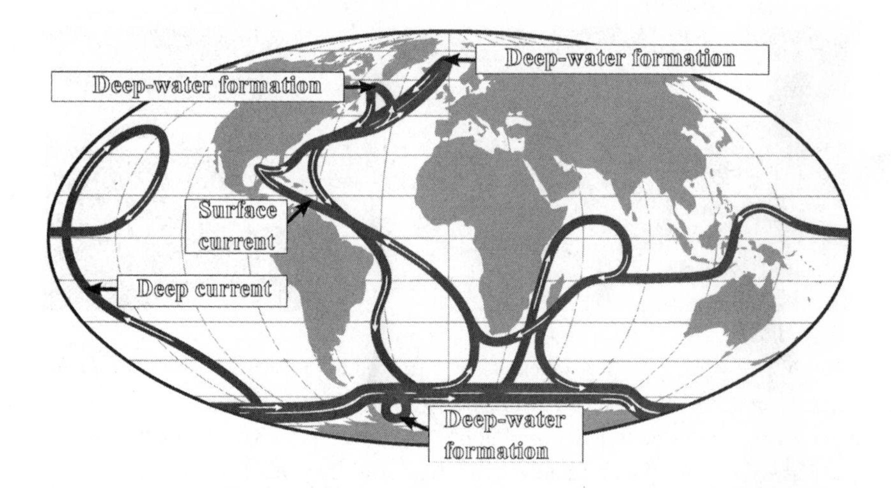

Figure 2.12 The global conveyor belt. Cold water flows deep, underneath the warmer surface currents. Areas labeled "deep-water formation" are the engine rooms, where salinity and temperature changes drive the currents, as explained in the text. (Adapted from an image in NASA, "Explaining Rapid Climate Change: Tales from the Ice," Earth Observatory, http://earthobservatory.nasa.gov/Features/Paleoclimatology_Evidence/paleoclimatology_evidence_2.php)

Heat is transported around the global conveyor belt at the rate of 1.2×10^{15} watts—that is, 1.2 billion megawatts (as measured at 24°N in the North Atlantic). Needless to say, this enormous heat transfer greatly influences the world's weather and climate. For example, the temperature in the North Atlantic is perhaps 5°C (9°F) higher than that in the North Pacific at similar latitudes; this difference may well be due to heat being delivered to the North Atlantic by the global conveyor belt. Significantly, we see from figure 2.12 that deep-water formation occurs in the North Atlantic but not in the North Pacific (due to reduced salinity of the Pacific, which is likely because of its higher freshwater inflow).

Global warming may slow down the global conveyor belt; the consequences of such a slowing have different effects on climate in different parts of the world, as we will see in chapter 4. Melting polar ice reduces the salinity and density of water at the polar deep-water formation zones, thus reducing the conveyor belt driving force. Meddies are thought to help here, by adding extra salt from the Mediterranean to the warm water that is heading toward the Arctic. Thus meddies counter the effect of melting polar ice in the North Atlantic, at least in terms of the global conveyor belt deep-water formation in this region.[13]

El Niño–Southern Oscillation

The El Niño–Southern Oscillation (ENSO) phenomenon is a good example with which to end this chapter, in that it illustrates two important points about ocean circulations:

1. They interact with the atmosphere.
2. They can have significant consequences for global weather.

One other aspect of ENSO is less satisfactory and yet perhaps it, too, is fitting for the ending of a chapter on this subject: the phenomenon is not well understood. Most of our knowledge of ENSO is empirical—it comes from observations. Much effort has been put into developing a theoretical understanding of the complex interactions between EN, the ocean part of the phenomenon, and SO, the atmospheric part. This effort has been directed toward improving our ability to predict El Niño years because the economic and human consequences can be disastrous if not mitigated by forewarning.

El Niño is a quasiperiodic temperature phenomenon that occurs in the tropical Pacific Ocean. It has been observed for a couple of centuries by fishermen that the tropical waters off the western coast of South America become unusually warm around the end of December in some years, now dubbed El Niño years. In other years, they are cooler than usual—La Niña years.[14] Underwater currents travel westward during El Niño and eastward during La Niña. The water temperatures oscillate between warm and cool, but the cycle is rather uneven and ragged. Thus El Niño happens every two to seven years, averaging twice per decade. The sea surface temperature rise averages 2°C (3.6°F). For a regular cycle, La Niña should follow El Niño as reliably as a trailer follows a truck; in fact, La Niña shows up only half as often and is seen as a sea-surface temperature drop of 1° to 2°C (1.8° to 3.6°F). Interestingly, the temperature changes of the ocean layer beneath the surface are greater than the surface temperature changes: 3° to 6°C (5.4° to 10.8°F) for El Niño and 2° to 4°C (3.6° to 7.2°F) for La Niña.[15] The duration of these changes is variable, averaging perhaps one year. The cycle is one over space as well as time so that warming at the western end of the equatorial Pacific Ocean correlates with cooling at the eastern end. The oscillation of warm water across the Pacific has been loosely compared with water sloshing across a bathtub.

The link between these temperature changes and weather appears in the form of cyclical changes in air pressure—the Southern Oscillation. A cool ocean surface is associated with high atmospheric pressure, and a warm surface with lower pressure. During an El Niño year, there is increased rainfall and reduced air pressure across the eastern equatorial Pacific and reduced rainfall and increased air pressure occurring across the western equatorial Pacific, Indonesia, and Australia. Westerly winds arise in this western region. The tropical North Atlantic is subjected to reduced hurricane activity, August to October, with increased activity in the eastern North Pacific. The opposite effects occur during La Niña years. ENSO influences large numbers of people by subjecting them to droughts or floods, heavy rainfall, landslides, hurricanes, and economic disruption (for example, fish stocks move out of the warm waters during El Niño to more nutrient-rich cooler waters). Hence there is a need to predict El Niño. The strength of these effects varies widely, with 1982/1983 (2,000 deaths and $13 billion damage) and 1997/1998 (23,000 deaths and $32 billion damage) being particularly strong.

Not everybody loses: the Pacific Northwest of North America enjoys warmer and drier winters than normal during El Niño, as does eastern Canada. Southeast Asia and Central Africa enjoy drier than normal winters during La Niña years. You see from these data that weather across the world is influenced by temperatures underneath the tropical Pacific Ocean.

Researchers do not yet have a good understanding of how El Niño forms. The jury is out on how global warming is affecting it. Despite this lack of basic understanding, there exists some ability to predict El Niño years based on observations across the tropical Pacific of water temperatures and air pressures. Yet initial predictions were for 2014/2015 to be a weak El Niño year—it turned out to be very strong. Was is strong because of climate change? It is possible—the El Niño circulation exchanges heat between atmosphere and deep ocean waters, which provides a link with climate change—but climate model predictions are not yet good enough to make confident predictions about this very complex phenomenon.[16]

***

Electromagnetic power from the sun heats Earth variably due to rotation and to differing albedos. This differential heating leads directly to weather phenomena. Climate changes naturally over long periods of time due to small and slow, but significant, long-term oscillations in Earth's orbital parameters. The water cycle and the fast carbon cycle influence weather mainly by means of three atmospheric greenhouse gases: water vapor, carbon dioxide, and methane. Ocean circulations, both shallow and deep, have an enormous effect on our weather and climate, because of the vast amount of heat they transfer about the planet. These circulations influence, and are influenced by, the atmosphere.

3

The Air We Breathe

The substance of the winds is too thin for human eyes, their written language is too difficult for human minds, and their spoken language mostly too faint for the ears.

John Muir

There is an intimate connection between atmospheric composition and weather. Here we tease out this connection by means of the science that implements it. The flow of air is the flow of heat, just as the flow of ocean currents serves to transfer heat. Indeed, the flow of air gives rise to some ocean currents, and heat differences in the oceans give rise to winds. There is a complex atmosphere–ocean feedback system (made more complex by the influence of landmasses) with many moving parts. The moving parts of the atmosphere are what we call weather. We metaphorically plunged into the ocean in chapter 2; here we soar into the air.

Composition and Structure

Around 3.4 billion years ago, outgassing from volcanoes gave us the nitrogen that today makes up 78.08% of our atmosphere. Around 2.4 billion years ago, life proliferated and contributed oxygen, which today makes up 20.95% of our atmosphere. The third most common gas that we breathe is argon, a noble gas—meaning that, like helium and neon, it is chemically inert—which leaks out of the ground and constitutes 0.93% of our atmosphere (and is also a fairly common constituent of seawater). The next most common gas is water vapor, which evaporates from surface water and transpires from the leaves of plants. Unlike the other gases, water vapor is confined to the lowest layer of the atmosphere (box 3.1). Also unlike that of the other constituents, the abundance of water vapor is variable. By mass, the average is about one-quarter of 1%; by volume about 1%, though it varies from virtually zero in the air above deserts and cold polar regions to 4% above tropical rain forests.

Box 3.1
Latent Heat

More than 99% of the water that resides in our atmosphere is in the lowest level, the troposphere. About 99% of this water is in the gaseous phase—water vapor. The remainder is liquid water droplets in clouds, rain, or fog. Water vapor is significant in meteorology for several reasons, the most important of which is that it transports vast amounts of heat energy around the atmosphere. There are two types of heat: sensible (heat we can feel) and latent (heat locked up in the molecular structure). Latent heat dominates and is a major determinant of surface meteorology; for example, it is responsible, as we will see, for powering up hurricanes. The important parameter here is the *latent heat of vaporization*, which tells us, for any given substance, the amount of energy it takes to convert the substance from liquid form into a gas. For each kilogram of water at standard (sea level) atmospheric pressure, this parameter is 2,260 kilojoules.

Here is a simple process that will help fix in our minds the relative importance of latent heat. Consider heating 1 kilogram (2.2 pounds) of water until it all evaporates as steam. We can calculate how much heat energy this takes. The water first heats up from room temperature (15°C [59°F]) to boiling point (100°C [212°F] at sea-level atmospheric pressure) and then it changes phase, becoming steam at the same temperature. Of course, water can evaporate at temperatures below the boiling point, but at the boiling point it *must* evaporate, by definition, if heating continues. Many substances have this property (a maximum temperature at which the liquid phase can exist for a given pressure). It takes 356 kilojoules of heat to raise 1 kilogram of water from 15°C to 100°C. This energy is the *sensible heat*—we can certainly feel the difference. It takes a further 2,260 kilojoules to convert the liquid water at 100°C into steam (that is, water vapor) at 100°C. This is the *latent heat*. Thus we invested a total of 2,616 kilojoules of energy into heating the water, but we feel only 356 kilojoules of it—the rest resides in the vapor.

When water vapor condenses, it releases the latent heat. If this happens in a cloud, then the air in the vicinity of the cloud warms up. We saw in chapter 2 that 500,000 cubic kilometers (121,000 cubic miles) of precipitation falls each year on the surface of Earth, and so, assuming that the average atmospheric water vapor content does not change much, year to year, this amount of water must evaporate from the surface. Doing the math, we find that 1,130,000,000,000,000,000 kilojoules of energy are transported to the atmosphere each year, as latent heat. This rate of energy transfer corresponds to a power of 36,000 gigawatts, which is about twice the total power currently consumed by all of humanity.

The remaining gases that make up our air exist in trace amounts: carbon dioxide, methane, ozone, nitrous oxide, hydrogen, and helium and other noble gases. Of course, we now know that some of these minor constituents—the first four in my list, beginning with carbon dioxide (at 0.04%, the most common trace gas)—have a significance for our weather and climate that is disproportional to their abundance, because they are greenhouse gases. We will soon see how and why they (plus water vapor, the most common and important greenhouse gas) matter.

This mixture of gases—this air of ours—influences climate and weather by absorbing incoming shortwave solar radiation and reemitting it as longwave radiation. The degree to which radiation is absorbed depends on frequency (the *absorption spectrum*) and is different for different gases. So air with a different mixture of gases would have a different absorption spectrum and so may have a different climate. The mixture of gases that make up the atmosphere has changed over the eons and climate has most certainly changed with it. However, the atmospheric constituents have been pretty stable for the last 200 million years, during which period the climate has nevertheless changed many times, and so atmospheric constitution is, unsurprisingly, not the whole story.

Apart from water vapor and ozone (see later) the gases are quite evenly distributed throughout the atmosphere, except for the outermost layer, which is of little concern to us as it is very tenuous and has no influence on weather. The layers of our atmosphere are, from the bottom up,

- Troposphere (0–12 km altitude)
- Stratosphere (12–50 km)
- Mesosphere (50–80 km)
- Thermosphere (80–700 km)
- Exosphere (greater than 700 km)

The troposphere contains almost all the atmospheric water and weather; indeed, it contains most of the air, even though it is a thin layer. Basic physics leads to the theoretical conclusion (quite closely reflected in reality) that the density of air falls exponentially with increasing altitude. More precisely, air density drops by half for every 5.5-kilometer (3.4-mile) increase in altitude above Earth's surface. This means that 78% of the mass of the atmosphere lies below the *tropopause* (the upper boundary of the troposphere); 99.8% lies in the troposphere and stratosphere. Thus these layers contain almost all the heat of the atmosphere and so, for our purposes, are the only important layers.

The atmosphere weighs more than 5,000,000,000,000,000 tons. One-quarter of 1% of this amount is a shade under 13 trillion tons, so this is the average weight of water that resides in the troposphere. The air naturally cools with increasing altitude above the surface, in part because this atmospheric water absorbs the longwave radiation emitted from the surface. In fact, however, the drop in temperature is not exponential like the drop in density; observations show that near the surface, ambient air temperature falls linearly with height above the surface at an average rate of approximately 6.5°C (43.7°F) per kilometer. Meteorologists refer to this temperature gradient as the *environmental lapse rate*; we will see in chapter 8 why it is a very important factor in determining the stability of air and so in determining how air moves when heated.[1]

I noted that the average tropopause altitude is 12 kilometers (7.5 miles). It is important to bear in mind that this value changes with latitude (also with seasons, being higher in summer than in winter, outside the tropics). At the equator, the troposphere reaches 18 kilometers (11 miles) into the sky but only 8 kilometers (5 miles) at the poles.

It defines the boundary between troposphere and stratosphere. The stratosphere is characterized by a temperature that *increases* with altitude (figure 3.1). How does this profile, so different from that of the troposphere, arise?

Most of the ozone (O_3, an uncommon form of atmospheric oxygen) is in the stratosphere. Ozone absorbs ultraviolet (UV) radiation from the sun,[2] and so its main source of heat is from above, not below—hence the reversed profile. This behavior—temperature rising with altitude—is known as a *temperature inversion*, and it results in very stable air, as we will see in chapter 8. In practice this means that air from the troposphere and that from the stratosphere do not mix much. Thus we can regard the tropopause as something of a glass ceiling as far as rising air is concerned: warm air will rise but not usually beyond the tropopause.[3] Clouds form at various altitudes but

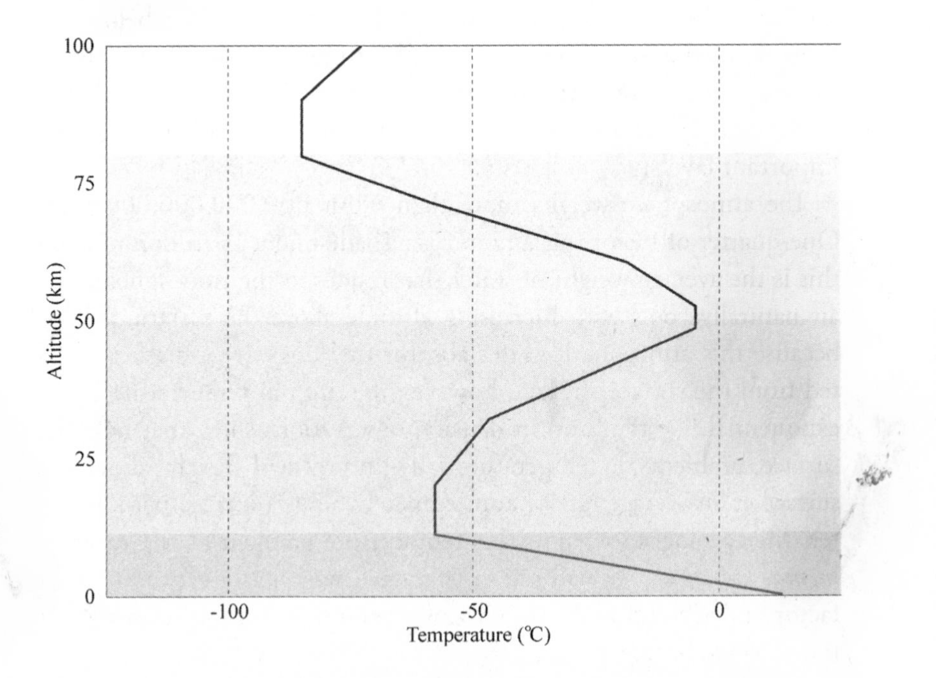

Figure 3.1 Atmospheric temperature profile: temperature (degrees centigrade) versus altitude (kilometers). Shading displays the boundaries between the four lowest layers of the atmosphere: troposphere, stratosphere, mesosphere, and thermosphere. Note the linear decrease of temperature with height in the troposphere and the increase in temperature in the stratosphere.

most stop at the tropopause. The boundary between troposphere and stratosphere also contains the *jet streams* (high-speed currents of air), as we will soon see.

Above the stratosphere, we lose interest—at least as far as weather is concerned. The mesosphere is the coldest layer of the atmosphere. The thermosphere is where the auroras appear; but the aurora borealis and its southern counterpart, the aurora australis, are part of space weather and not the terrestrial weather than forms the subject of this book. (The lower part of the thermosphere is ionized and reflects radio waves and thus is of sufficient importance to humans to be given its own name; it is known as the ionosphere.) The exosphere is what is left of the atmosphere above the thermosphere; air molecules in the exosphere are kilometers apart and the space between them is, well, space. To emphasize the increasingly tenuous nature of the air at these higher altitudes, let me note that space officially begins at the Kármán line, 100 kilometers (60 miles) above Earth's surface.

Absorption and Emission

The mix of gases in our atmosphere determines its absorption spectrum. The important measure of this spectrum from our perspective is the opacity of the atmosphere to electromagnetic (EM) radiation as a function of radiation wavelength (figure 3.2). The opacity is the sum of all

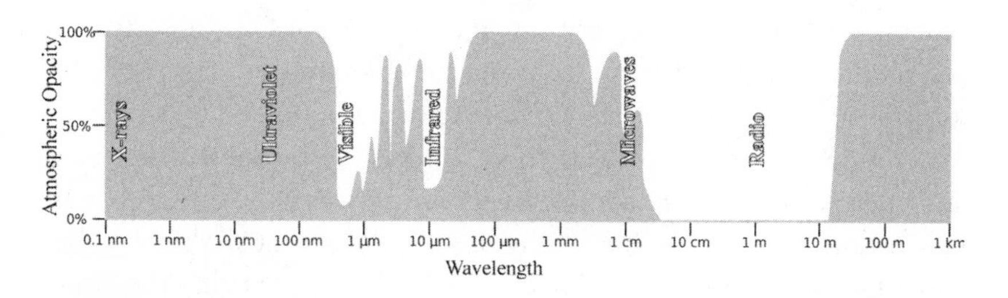

Figure 3.2 Atmospheric opacity to electromagnetic radiation versus radiation wavelength. Zero opacity means that the atmosphere is totally transparent; 100% opacity means that it is totally opaque—it scatters or absorbs all radiation. (Adapted from an image in NASA, "Atmospheric Electromagnetic Transmittance or Opacity," Wikimedia Commons, https://commons.wikimedia.org/wiki/File:Atmospheric_electromagnetic_transmittance_or_opacity.jpg)

the differing opacities of the different atmospheric constituents. Most radiation is absorbed by water vapor, but there is significant absorption from carbon dioxide, from other greenhouse gases such as ozone, and from oxygen. Note that UV radiation and shorter wavelengths (more energetic, and therefore more damaging to life) such as X-rays are more or less completely absorbed before they reach the surface. Visible light is the shortest-wavelength radiation to reach the surface in significant amounts, which presumably is why eyes are sensitive to it.[4]

The shortwave EM radiation reaching the surface of Earth differs from the radiation reaching the top of the atmosphere in part because of albedo (some radiation is reflected) and in part because of absorption, as we saw in figure 2.3. Absorption is responsible for the frequency or wavelength dependence of radiation penetration and so for the greenhouse effect. This greenhouse phenomenon was introduced in chapter 1; here we examine it in more detail.

The power density of shortwave radiation reaching the top of the atmosphere is, on average, 341 W m^{-2}; we derived this figure from the solar flux data in chapter 2. It is worth emphasizing that this figure is a global average; it varies greatly with time of day and quite a lot with latitude and season. It also varies with cloud cover, as you can see from figure 3.3. Note that 102 W m^{-2} of this total are reflected off clouds and Earth's surface back into space. The remainder is absorbed by the atmosphere or surface and is then reradiated according to the blackbody law described in chapter 1.

The global average longwave radiation budget of the surface and the atmosphere is illustrated in figure 3.4. Air molecules absorb some shortwave radiation directly from the sun and absorb longwave radiation emitted by Earth's surface. More energy is absorbed by the atmosphere via convection of heat from the surface transferred by rising air (thermals), and as latent heat held within water molecules that are taken aloft on rising air. Blackbody theory tells us that all this absorbed radiation must be reemitted by the atmosphere, given that the atmosphere is more or less at thermal equilibrium (ignoring, for now, global warming). The longwave emissions go into space or are absorbed by the surface (the "back radiation" of figure 3.4). Do the math, and you can see that the budget for energy absorbed by, and reemitted by, the atmosphere is nearly in balance. The same is true for the surface if you remember to include the 161 W m^{-2} of shortwave radiation power that

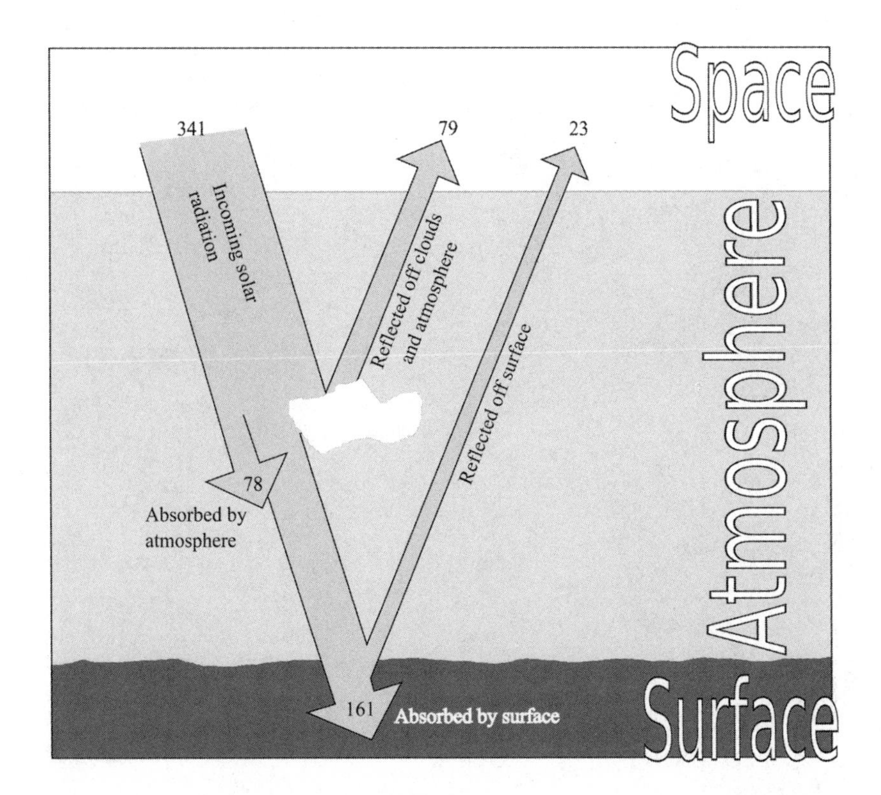

Figure 3.3 Average shortwave radiation (numbers, W m^{-2}) absorbed and scattered by the atmosphere and by Earth's surface. (Data from K. E. Trenberth, J. T. Fasullo, and J. Kiehl, "Earth's Global Energy Budget," *Bulletin of the American Meteorological Society* 90 [2009]: 311–323)

is absorbed (see figure 3.3). The difference between incoming and outgoing radiation power is about 1 W m^{-2}; this difference is what we call global warming. Of course it is an average value; given the variation of surface albedo and of atmospheric opacity (variations in time and with location) it is not hard to imagine the difficulty in obtaining the numbers in figures 3.3 and 3.4 by detailed observation. Nevertheless, data are now unequivocal: global warming is happening. We will investigate this phenomenon in more detail in chapter 4; here the (near) balance is more noteworthy, as it indicates that the surface of Earth and its layered atmosphere are almost at thermal equilibrium with each other and with the rest of the universe—energy input is (very nearly) matched by energy output.[5]

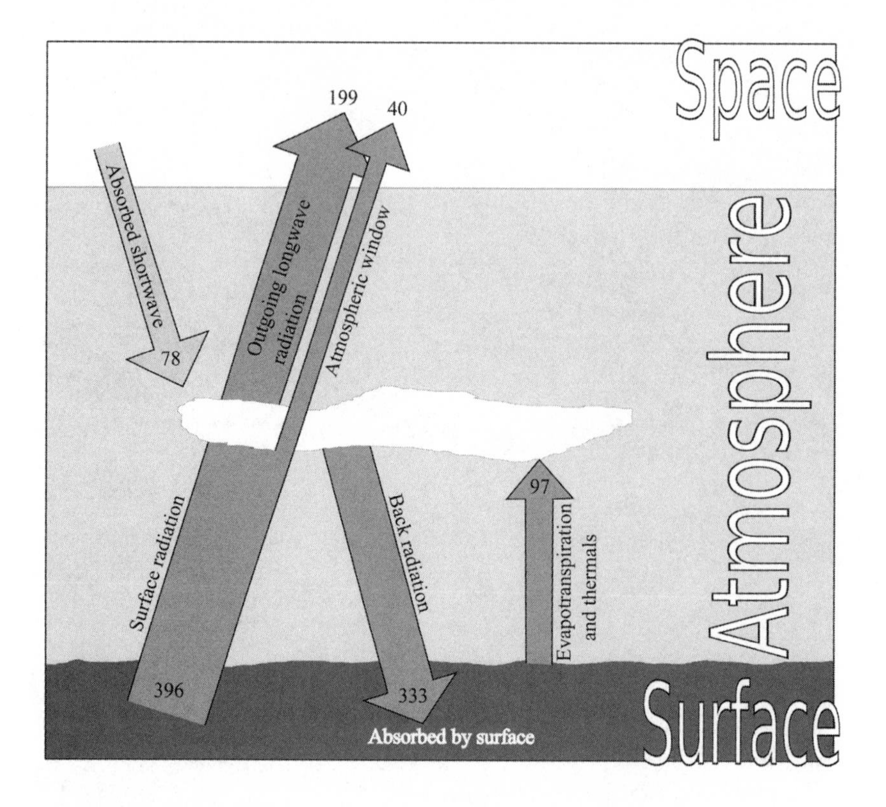

Figure 3.4 Average longwave radiation (numbers, W m^{-2}) absorbed and scattered by the atmosphere and by Earth's surface. The 78 W m^{-2} of shortwave radiation absorbed by the atmosphere (see figure 3.3) is also shown, so readers can verify that the atmospheric power budget is nearly balanced. The atmospheric window is a gap in the absorption spectrum through which longwave radiation can leak from the surface into space. (Data from Trenberth, Fasullo, and Kiehl, "Earth's Global Energy Budget")

Tropospheric Circulations

Differential heating of the air results in atmospheric circulation just as differential heating of the ocean results in ocean currents, as we saw in chapter 2. Indeed, we noted that the two fluids—ocean and atmosphere—were intimately connected. One connection that is hugely important for weather physics is that the atmosphere transfers 2,200,000,000,000,000 watts of power from the world's oceans to the land, averaged over a year (mostly during the northern winter). That's about 15 W m^{-2}. There are similarities between oceanic and atmospheric flows: both transmit solar heat from the equator poleward (of similar magnitudes, with the atmosphere coming out on top, as it were); both are much influenced by the Coriolis force; both are layered. There are differences: unlike air currents, ocean currents are strongly influenced by surface topography;[6] oceans, being liquid, do not vary in density nearly so much as atmospheric air and do not contain water vapor.

Cyclones and Anticyclones

Large-scale atmospheric circulations begin at the equator and, more weakly, at the poles. Let us begin in the tropics. Because equatorial regions receive a higher solar power density than the rest of the world's surface, the equatorial air gets hotter and so rises. In fact, it often rises as high as it can go—to the top of the troposphere (the tropopause: a glass ceiling, you may recall). This is tropical convection on a global scale.

The manner in which air rises and falls on a somewhat smaller scale (but still large enough for the Coriolis force to be significant) is illustrated schematically in figure 3.5. Rising air creates a low-pressure center that pulls in nearby warm surface air (see figure 3.5*a*). The Coriolis force causes this moving air to veer to the right (in the Northern Hemisphere), and so the incoming air creates a spiral. This dynamic system is called a *cyclone*.[7] The situation is similar for sinking air, with the directions reversed (see figure 3.5*b*). Cool air aloft sinks, creating a high-pressure zone near the surface; air spilling out from this center veers to the right, again creating a spiral—an *anticyclone*. Think of bath water draining out of the plug hole. These cyclones and anticyclones are the mid-latitude highs and lows on television weather maps. These are what we mean when we talk about rising or sinking air; (anti)cyclones are the means by which air changes altitude in the atmosphere on this synoptic scale (of the order 1,000 kilometers [620 miles]).

Our main focus in this chapter is on the first of these phenomena: the planetary-scale movements of air, which begin with tropical convection. It is to these global elevators or conveyor belts of air movement that we now turn.

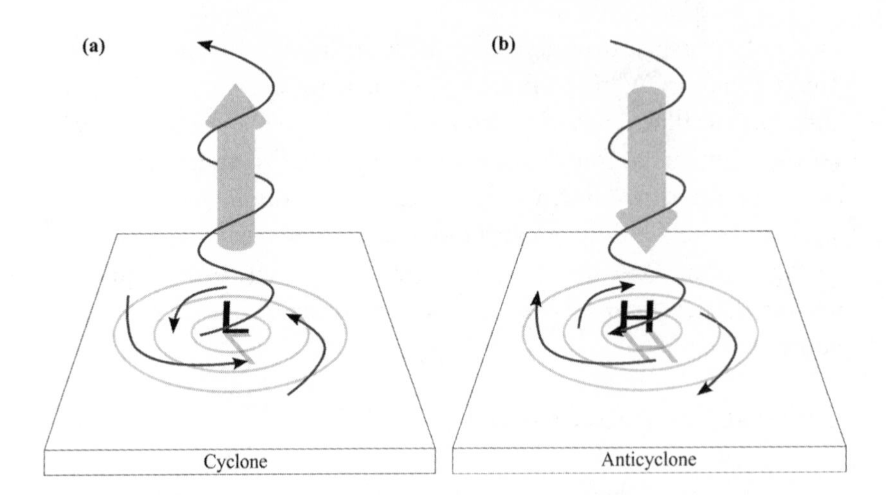

Figure 3.5 Cyclones and anticyclones. (*a*) Rising air creates a low-pressure center (L, with isobars shown as *gray circles*). For a large-scale system, hundreds of kilometers across, surface air approaching the central low is turned to the right (if north of the equator) by the Coriolis force, resulting in the spiral motion shown. (The wind flow would be geostrophic if the flow lines were parallel to the isobars.) (*b*) Falling air results in a high-pressure center and the opposite rotation. These large-scale cyclones and anticyclones are the means by which air rises and falls in global circulations patterns.

Hadley Cells

Most of the rising air in the atmosphere is to be found near the equator at the *Intertropical Convergence Zone* (ITCZ). Here warm, moist air over the oceans rises high into the atmosphere in a series of cyclones around the world. The air cools as it rises and its water vapor condenses, forming clouds and often thunderstorms. The detailed mechanism is discussed in chapter 7. A large amount of rain falls in the ITCZ, hence tropical rain forests (figure 3.6*a*).[8] The rising air is replaced by warm surface air to the north and south that heads toward the equator. As the warm air moves, the ubiquitous Coriolis force deflects it to the right (westward) in the north and to the left (also westward) in the south, resulting in the northern and southern trade winds, as indicated in figure 3.6*b*. (There is no Coriolis force right at the equator and this region is often squally, with slow and irregular winds; it was known in the Age of Sail as "the doldrums," where ships could be becalmed.)

(a)

(b)

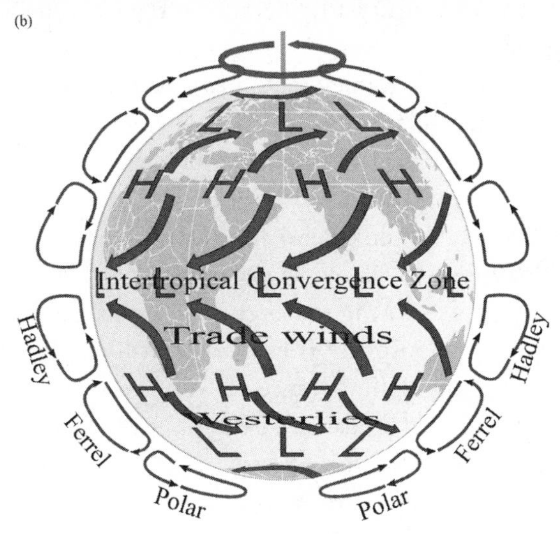

Figure 3.6 Atmospheric circulation. (*a*) The Intertropical Convergence Zone is marked by a line of clouds in this satellite image. (*b*) The Hadley cells, Ferrel cells, and polar cells are doughnut-shaped circulation patterns that girdle Earth, rotating as indicated. They transport heat from the tropics to the poles. The Hadley and polar cells are driven by temperature differences; the Ferrel cells that lie in between are driven by heat transfer arising from the frequent and widespread mid-latitude storms. The direction of circulation creates latitudes of high and low surface air pressures; add the Coriolis force, and we see why the northern and southern trade winds exist, and the westerlies and the polar easterlies (*not shown*). The winds shown (*arrows*) are at the surface; up near the tropopause, the wind direction is generally opposite—though not at mid-latitudes. There is heavy rain at the ITCZ, near the equator, and there are deserts near the Hadley–Ferrel boundaries. ([*a*] From NASA, "The Intertropical Convergence Zone," Earth Observatory, http://earthobservatory.nasa.gov/IOTD/view.php?id=703)

So the warm air that rises in the ITCZ cools, and is pushed away from the equator in both directions at high altitudes near the tropopause. Farther north and south, it finds itself over warmer, less dense air and so sinks, completing the circulation pattern. The Coriolis force pushes the winds westward, and so the circulation is not purely north–south (see figure 3.6*b*); the wind trajectory is actually helical: giant corkscrews heading westward around the world. The component of this complex trajectory that is just north and south of the equator is a doughnut-shaped ring of air that girdles Earth, called a *Hadley cell*, after the eighteenth-century English meteorologist who first appreciated its existence.

The explanation for Hadley cells just given does not tell us the northern and southern latitudes at which the moving air sinks; that is, it does not tell us how far from the equator these cells extend. In fact, they stretch to about 30° north and south. If Earth were not rotating, then the Hadley cells would extend almost to the poles, but on a rotating Earth this cannot be. Why? Consider the northern cell only (the same reasoning applies to the southern cell, with directions reversed). Air high up in the troposphere is heading north, cooling all the time. It is also veering to the right, because of—well, you know by now. You may recall that one of the characteristics of the Coriolis force is that it gets stronger away from the equator. Thus the farther north these winds get, the faster they move eastward. At high speeds, the air flow becomes unstable and degenerates from smooth flow into eddies. On our Earth, today, the instability occurs in the vicinity of 30°N, marking the northern extent of the Hadley cell.[9] Another way to understand why Hadley cells do not persist to the poles is to consider again the poleward-moving air that is high up near the tropopause. This air becomes bunched together as it moves poleward (as lines of longitude converge) and so sinks due to increased density.

The descending air at 30° latitude, north and south, lost its water content back at the ITCZ and so the regions at 30° latitude are characterized by dry air as well as high pressures—hence many of the world's deserts are located nearby. Historically known as "horse latitudes," these are regions of high pressure and variable winds or calm.[10] Observations over the past 32 years have shown a statistically significant shift in the 30° extent of Hadley cells to higher latitudes. The Hadley cell system (north and south cells taken together) is growing bigger. It is also growing more powerful: recent measurements show that this system generates atmospheric kinetic energy (the energy due to movement) at a rate of 198 trillion watts, and that this rate is increasing by 540 billion watts per year. The extra energy is due to global warming.

More Cells

Polar cells are giant atmospheric convection cells located in—you guessed it—polar regions, driven by temperature differences (think of the circulation patterns you see in your latte). In this sense, they are like the much larger and more powerful Hadley cells. Their influence on energy transfer is also the same: they move heat toward the poles. Air high up above the North Pole (say; the same happens at the South Pole) becomes cold and dense. It sinks (creating a high-pressure zone at the pole) and spreads southward along the surface, warming as it moves into temperate latitudes. At about 60°N, it rises and is pulled back toward the pole by the pressure gradient aloft. The Coriolis force is particularly strong in this region, and the north–south movements are strongly deflected, creating the polar easterlies (arrows near the poles in figure 3.6*b*).

Ferrel cells, named after a nineteenth-century American schoolteacher, occupy the latitudes between Hadley and polar cells—that is, 30° to 60° north and south. Ferrel cells rotate in the opposite sense, as indicated in figure 3.6*b*. They are not driven by latitudinal temperature differences as are the other cells; instead, they seem to be a response to the heat transport of those ubiquitous mid-latitude storms. Measurements show that Ferrel cells need 277 trillion watts of power to generate the kinetic energy of air movement that defines them.

More than in the other cells, air movement in Ferrel cells is complicated. The simple circulation shown in figure 3.6*b* is deceptive—it is an average along the north–south axis, hardly discernible in real time compared with other movements. There is also an easterly component to wind velocity near the surface (westerly aloft) attributable to the Coriolis force, but more complex are the vortices—eddies—that are characteristic of these latitudes. These are the lows and highs—circulations that are directly responsible for the unsettled weather of, say, the eastern seaboard of the United States or anywhere in Britain. They arise from the mixing of air in the Ferrel cells and in particular from the extensive meanderings of the polar jet stream, discussed in the next section. Ferrel cells also transport heat in the direction opposite that of the Hadley and polar cells (that is, toward the equator). The net effect of these atmospheric circulation cells is to carry 6,000,000,000,000,000 watts of power to the poles. (This heat is partly sensible—that is, discernible heat, or warm air—but is mostly latent heat locked up in water vapor molecules.)

The three-cell model—Hadley, Ferrel, and polar—is intended to describe the average global atmospheric circulation. In this sense it is a model of climate rather than of weather, though it clearly is instrumental in determining local weather conditions (we're at the equator: it's going to rain). The averaging process represents a simplification for all the cells, not just for Ferrel cells: please do not take the circulation diagram too literally (see figure 3.6*b*); it applies, but at any instant of time a snapshot image of wind velocities will resemble this diagram only partly and only in certain places. The trade winds are fairly reliable, as are equatorial rains, but the circulations are variable, sometimes discernible only when averaged as in the Ferrel cells. Fluid flow is complex even with the simplest boundary conditions (say flow along a fixed channel with smooth and regular boundaries); here we have irregular boundaries due to topography on a spheroid that is spinning. This is messy.

We have three cells in each hemisphere. Other planets have more, or fewer. Earth likely has had more or fewer in the past, depending on temperature gradients and land distribution. The three-cell model is not immutable. It is also not perfect. Although a good description of surface wind distributions, and of tropical circulation patterns, it fails to describe very well the upper tropospheric circulations in temperate regions. It is, like much of our subject, a work in progress.[11]

Jet Streams

Familiar, at least in name, from your daily weather reports (particularly in winter), jet streams are high-altitude ribbons of air moving eastward at high speeds. The ribbons are typically thousands of kilometers in length and hundreds of kilometers wide but only a few kilometers deep. There are four main jet streams; at any given instant there may be several others, but the four most important jet streams are what we will concentrate on here.[12]

Our main jet streams are located in latitude at the boundaries of the Ferrel cells—nominally 30° and 60° north and south—and their altitude is just below the tropopause. The "polar front jets," or *polar jets*, occur between the polar and Ferrel cells, typically 10 kilometers (6 miles) above the surface and with wind speeds of around 250 kilometers (155 miles) per hour. The *subtropical jets* are a little higher up, say 12 kilometers (7.5 miles) above the surface, and they move a little slower. The flow of jet streams is geostrophic. Subtropical jets are not as concentrated—they are more diffuse—than the polar jets, for reasons we will soon uncover. Both sets of jet streams influence surface weather; the polar jets are more significant because they are more powerful and because they are a little closer to the surface.[13]

The locations and wind speeds just given are typical values; it is worth emphasizing at the outset what will become abundantly clear over the next few paragraphs: that speed and (especially) latitude of jets are highly variable. They move with the seasons—poleward in summer and toward the equator in winter. Occasionally the subtropical and polar jets merge. Jets help to define the boundaries between atmospheric circulation cells and so it follows that the latitudes of these circulation cells also wander, sometimes far. The jets oscillate as they move around the globe, as suggested in figure 3.7*a*; these *Rossby waves* are rather long—there are usually four to six wavelengths per jet, though the number and amplitude change over timescales typical of large-scale weather phenomena—say a week.[14]

The origin of the jets is clear enough: they arise from a combination of horizontal temperature gradients and the Coriolis force. We have seen how equatorial temperatures set up poleward winds that are subjected to the Coriolis force; these become geostrophic when the force due to poleward pressure difference is matched by the Coriolis force. So these winds proceed in an eastward direction.

The polar jets are more concentrated and their location marks a sudden change in the height of the tropopause, as we see in figure 3.7*b*. These facts are related, as follows. Consider figure 3.8*a*. Cold, dense air from a polar cell intrudes beneath the warmer, lighter air of the contiguous Ferrel cell, creating the more or less permanent feature of these latitudes known as a *polar front*. This polar front is characterized on the surface by a large change in temperature over a short distance, as shown in the figure. Now consider the height of the tropopause—that

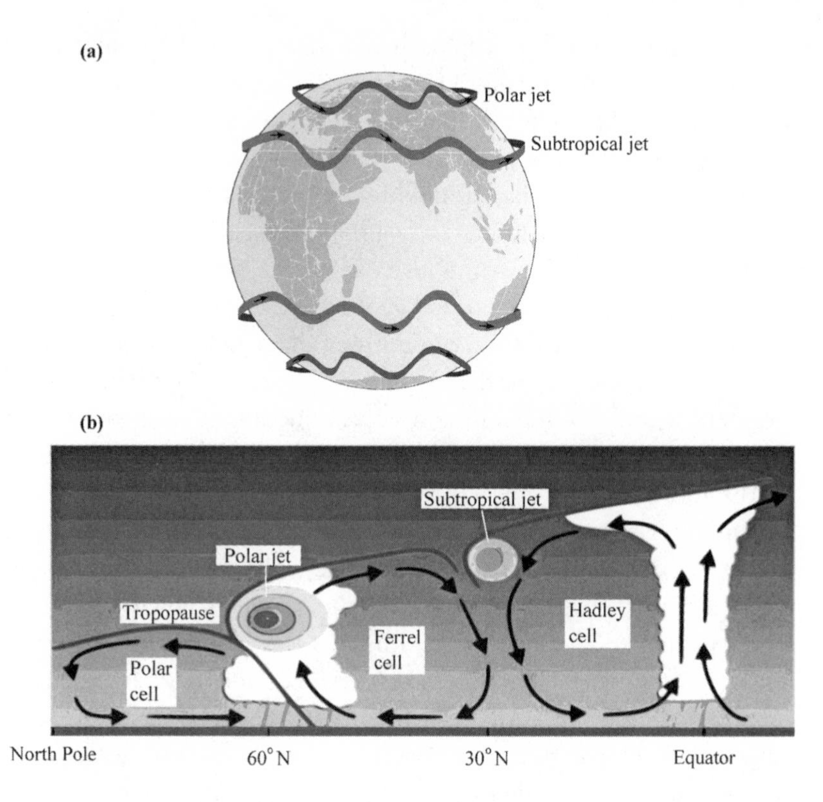

Figure 3.7 Jet streams. (*a*) The four main jet streams travel eastward at high altitude, just beneath the tropopause. The polar jets are narrower and faster than the subtropical jets. They are located at the boundaries between the main circulation cells, and their actual trajectory is very variable. (*b*) Side view of the Northern Hemisphere, showing the relative positions of the Hadley, Ferrel, and polar cells and the polar and subtropical jets. Note the rain-producing clouds at latitudes of low pressure, rising to the top of the troposphere. The jets are moving east, into the page. (From National Weather Service, "The Jet Stream," in "Global Weather," National Oceanic and Atmospheric Administration, http://www.srh.noaa.gov/jetstream/global/jet.html)

is, the vertical extent of the troposphere. The height at any given point on the tropopause increases as the average temperature below the point increases; the height decreases as the average temperature falls. (The reason for this behavior is that convective mixing increases with temperature and this mixing stirs up the troposphere.) The consequence of

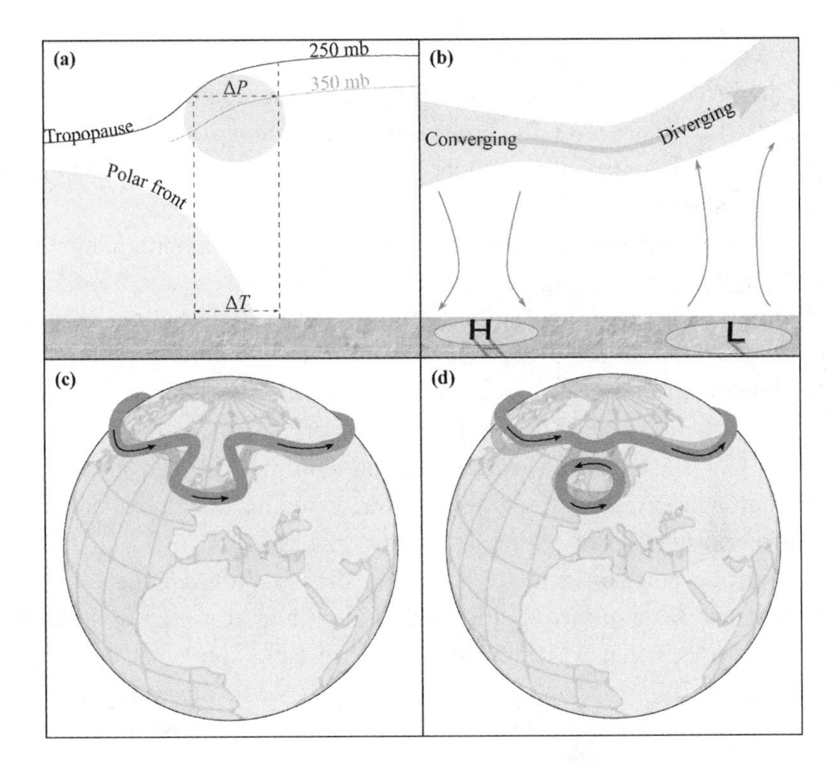

Figure 3.8 Polar jet. (*a*) A large temperature change ΔT at the surface converts into a large pressure gradient ΔP aloft—here we choose a representative value of 250 millibars (a quarter atmosphere) for pressure at the tropopause. This pressure gradient results in a poleward force on the air, which becomes geostrophic (turns into the page, here) due to the strong Coriolis force. The *gray ellipse* indicates polar jet stream, into the page. (*b*) A converging jet stream leads to dense, downward bursts of air that generate high-pressure areas. Similarly, low pressures at the surface result from a divergent jet stream above it. (*c*) The meander of the northern polar jet increases, causing (*d*) a cold-air bubble to pinch off from the jet, resulting in a depression in the Ferrel cell. Such vortices are typical of the turbulent Ferrel regions.

a polar front is therefore a rapid change in the height of the tropopause above the front, as indicated in figure 3.8*a*.

Indeed, sometimes the change is so rapid as to be discontinuous, as the tropopause "folds in" to the boundary separating polar and Ferrel circulation cells, as suggested in figure 3.7*b*. (In this case, stratospheric air becomes mixed up with the warmer tropospheric air.)

Now we can see why the polar jet is more concentrated and powerful than the subtropical jet. The large horizontal temperature gradient indicated in figure 3.8*a*, in combination with the rapid change in tropopause altitude, means that there is a large horizontal difference in air pressure aloft, as indicated.

This pressure difference pushes high-altitude air poleward, assuming that the polar front is aligned east–west, which it is, on average. The Coriolis force, which is stronger at the latitude of polar jets than at the latitude of subtropical jets, then deflects the air flow. The air flow—the polar jet—becomes geostrophic when it is heading east (into the page, for the case shown in figure 3.8*a*). The combination of large temperature gradients, strong Coriolis force, and reduced friction aloft gives us high-speed and concentrated jets. (Horizontal temperature gradients tend to be stronger in winter, which is why winter jet streams flow faster.)

An important consequence of this jet stream physics is the association of jets with surface weather below. Looking at a map of surface temperatures, you can pick out the polar front and know that the polar jet is above it. If you look at a map of isobars for this region, you will see the polar jet where the isobars run closest together. If your local weather forecast shows that the polar jet stream is passing over your head in the next few days, from north to south, then you know that you will be passing from Ferrel to polar cell (if you live north of the equator) and that colder weather is on its way.

Jet streams are three dimensional; they move up and down as well as around the world. As they move up and down, their density and cross section (area as measured perpendicular to flow direction) change. In figure 3.8*b*, you can see how a convergent and a divergent jet stream lead to vertical air flow and so to high- and low-pressure regions on the surface. In figure 3.8*c* and *d*, we see an example of the consequences that can arise from the horizontal Rossby waves: a bubble of cold (in this case) air is pinched off—becomes a self-contained vortex detached from the jet stream—generating a North Atlantic depression. This behavior is characteristic of the polar jet and leads to the unsettled weather that is typical at their latitudes. In figure 3.8*d*, the depression happens to be over England.

We can think of the meandering polar jets as dropping vortices into the Ferrel cells, like water drops falling into a stream. These vortices are the eddies discussed earlier, the high- and low-pressure zones that lead to unsettled, faster changing, and less predictable weather in temperate regions, compared with that at the equator and poles.

A particularly striking example of the interconnectedness of meteorological fluid flow arises when we consider the following observation: the subtropical jet stream position over the tropical Pacific Ocean is different in El Niño years than in other years. We have already seen that deep ocean currents contribute to El Niño; here we see that air flow high up in the troposphere is influenced by and influences the same weather phenomenon. Thus El Niño demonstrates that deep ocean currents and high-altitude air flows interact.[15]

***

The composition and layered structure of our atmosphere influence the amount and distribution of shortwave solar power that reaches the surface. The heat generated by this power influx is reradiated as longwave power or is transported around the atmosphere as latent heat in water vapor. Differential heating at the surface drives large-scale atmospheric circulations, resulting in three tropospheric circulation cells in each hemisphere that girdle the planet. The Coriolis force acts on circulating air to produce large-scale atmospheric movements such as trade winds and jet streams.

4

Dynamic Planet

I will praise the English climate till I die—even if I die of the English climate.
G. K. Chesterton

Earth's climate has been slowly changing since it first arose. In this chapter, we examine the geophysical reasons why this is so. We review the recent human contribution to climate change and focus on our progress in developing an understanding of this contribution and its consequences.

This is our first hint at the complexity—some would say "perversity"—of matters statistical. After all, if climate is average weather, then how can climate change? If we average weather over 30 years, say, then we get climate. But this running average can change over decades or centuries. Similarly we can average over a millennium, or over a million years, and get a different viewpoint of climate.[1] More of such statistical ruminations are given in chapters 5 and 6; here we gently introduce climate models. But you've been warned.

Greenhouses and Other Worlds

A good way to get a grip of the physics that underpins the greenhouse effect is to look at the physics of greenhouses. There is a clear analogy with the climate effect introduced in chapter 1; this analogy is useful if we bear in mind its limitations.

Let us assume for the moment that you live in a cold and sunny part of the world—say, a mountain desert region such as the Great Basin in the western United States or the Gobi Desert in Central Asia. You want to grow delicate tropical plants such as hibiscus, in winter, for reasons best known to yourself. You build a cloche (also called a "row cover") to shelter your delicate blooms from wind (figure 4.1*a*), and you install

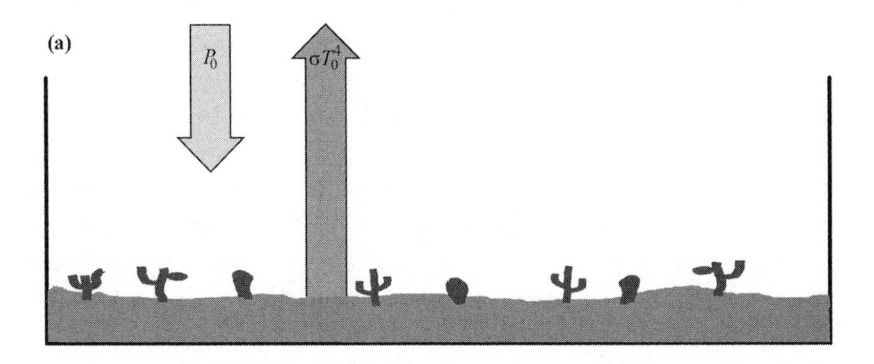

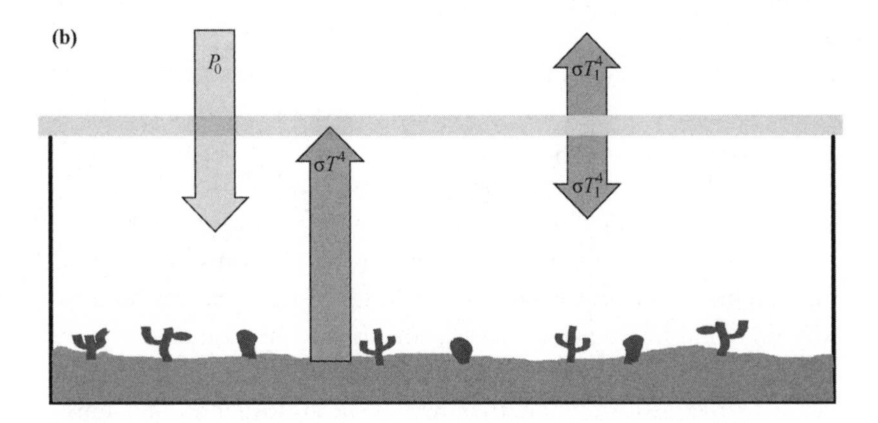

Figure 4.1 Idealized greenhouse. (*a*) Incoming shortwave solar radiation of power P_0 balances outgoing longwave surface radiation of power σT_0^4, assuming that the surface is a perfect blackbody in equilibrium, so we can calculate T_0 knowing only P_0. (*b*) Add a glass layer that is also assumed to be a perfect blackbody. Thermal equilibrium again means balancing power inputs and outputs so that $T_1^4 = T_0^4$ and $T^4 = 2T_0^4$.

a watering system. The only problem is ambient temperature, as we see by balancing the incoming shortwave solar power with the outgoing longwave blackbody power. With no glass atop the cloche, the average temperature is that of your local environment—let us say –3°C (27°F). This will play havoc with your hibiscus. Add glass, and the power balance changes (see figure 4.1*b*). We will assume for simplicity that glass absorbs none of the incoming shortwave radiation and all the outgoing longwave radiation—this is not too far from the truth. Assume also that the (nonreflecting) glass is a perfect blackbody at infrared wavelengths so that it radiates all the power it absorbs, but at longer wavelengths. Balancing the power and doing the math (here relegated to the appendix, for those who are interested), we find that the absolute temperature inside the cloche has increased by 19% to a toasty 48°C (118°F)—hibiscus heaven.

The analogy with the atmospheric greenhouse effect is as clear as the cloche glass. The glass plays the role of the atmosphere, passing most (here, all) of the incoming solar radiation yet absorbing the longwave radiation that is emitted by the surface. The glass, as well as the surface, is a blackbody and so radiates. To balance the power budget, note that the temperature of the glass has to be the same as that of the environment—here the analogy is misleading—whereas the interior temperature rises.[2]

It might seem natural that the glass/atmosphere temperature should be that for a blackbody that is emitting the same power as it absorbs in solar radiation. It happens to be so for the simple cloche calculation but is not the case for our atmosphere, as we saw in figures 3.3 and 3.4; the physics will be made clear soon enough with another simple power-balance example. In the calculation in the appendix it is also assumed that the glass radiates the same in both directions—there is no reason why it should not. Again, however, this is not the case for our atmosphere; we will soon see that the back radiation (to the ground [see figure 3.4]) constitutes more than half of the atmospheric emissions.

Suppose we add a second layer of glass to the cloche, above the first (but not in contact, so that we may regard them as separate blackbody radiators). A calculation similar to that in the appendix shows that the top layer of glass is at ambient temperature (–3°C [27°F]), the bottom layer of glass is at a temperature of 48°C (118°F), and the surface is at a temperature of 82°C (180°F). Again, we get the incorrect result (incorrect as an analogy for the atmospheric greenhouse) that the top layer—analogous to the top of the atmosphere—is at ambient temperature, but the correct result that the surface is warmer than this. In addition, we find that the lower layer of glass (the lower atmosphere) is at a temperature between the two—also correct.

Instead of a cloche or greenhouse, we might have picked a parked car for our analogy with the atmosphere/surface heating effect. There are sad and disturbingly frequent reports of fatalities that occur inside cars on a hot day with the windows rolled up—typically, they will be in a shopping-mall parking lot. The interior becomes so hot due to blackbody heating that vulnerable passengers—babies, pets—perish. The car analogy might be a better choice, not because of the extreme heating involved (though this certainly applies on über-greenhouse Venus, as discussed in the appendix), but because the car windows can be opened a little. This action vents hot air and mitigates the greenhouse heating effect. There is a broadly analogous open window operating in our atmosphere. In figure 4.2, we see part of the atmospheric absorption spectrum (it is a close-up of the spectrum shown in figure 3.2): note the wide gap in water vapor absorption at 8 to 15 micrometers (0.0003 to 0.0006 inch), partly filled in by carbon dioxide; blackbody radiation from the surface leaks through this gap as illustrated in figure 3.4 ("atmospheric window"). Of course, the analogy with a car window cracked open is a rough one: the window permits convective heat loss rather than radiative, and it is not limited to certain wavelengths.[3]

Before moving on to a more realistic model of the Earth–atmosphere power budget let us briefly consider another idealized example (figure 4.3). Suppose a planet has an atmosphere that is thin

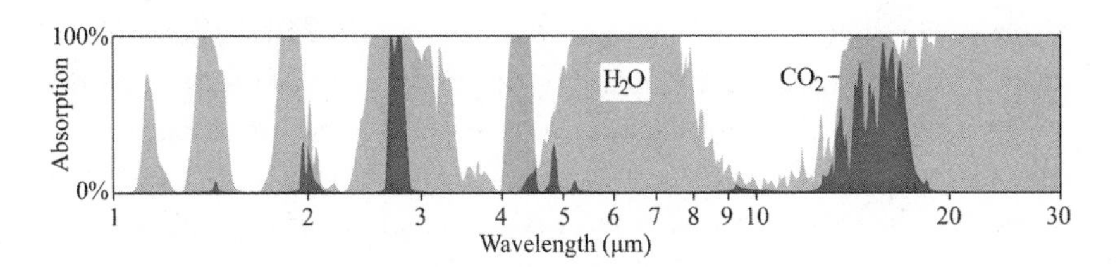

Figure 4.2 Atmospheric absorption spectrum for longwave radiation of wavelengths between 1 and 30 micrometers. Note the large window in the water-vapor spectrum between 8 and 15 micrometers; this window is partly closed by carbon dioxide absorption.

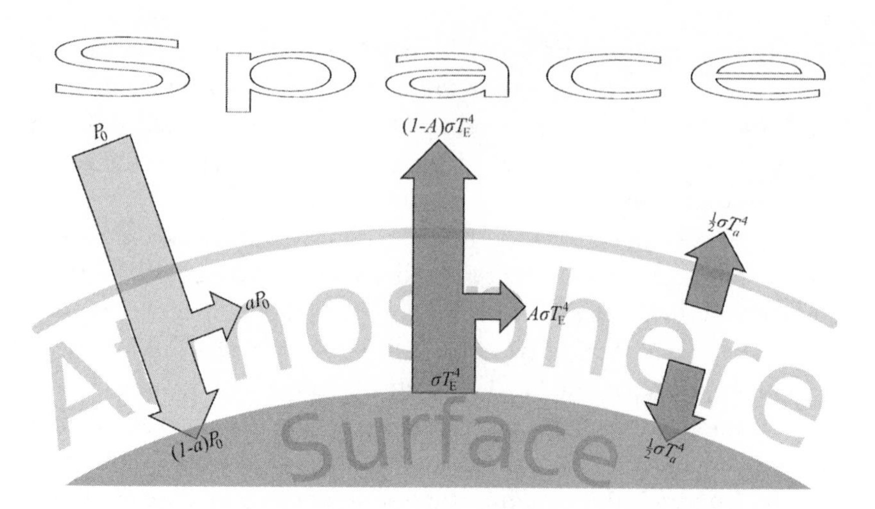

Figure 4.3 Power budget of a planet with a thin atmosphere. Incoming shortwave power P_0 flows to the surface and back again. The planet surface is taken to be a perfect blackbody, but the thin atmosphere is a gray body. A fraction *a* of shortwave (*A* of longwave) radiation is absorbed by the atmosphere. Because this atmosphere is thin, assume that it radiates equally up and down, as with the cloche glass in figure 4.1*b*. The calculations for this system are performed in the appendix.

so that we may regard its temperature as uniform, and that its surface and atmospheric structure are such that they reflect none of the electromagnetic (EM) radiation that arrives from its star. The surface is a perfect blackbody: it absorbs and reemits all the power it receives. The atmosphere is a so-called *gray body*: it absorbs only a fraction of the radiation that impinges on it. Say that it absorbs a fraction *a* of shortwave radiation from the star and a fraction *A* of longwave radiation from the surface. If this planet is in thermal equilibrium, then the power that enters the atmosphere must equal the power that leaves it, and the power budget of the surface is similarly balanced. (The math is shown in the appendix, for interested readers.) For such a planet, the surface temperature exceeds the blackbody temperature T_0 if *A* exceeds *a*, and the atmospheric temperature falls in the range $0.84T_0$ to $1.19T_0$, depending on the values of *a* and *A*. That is, the atmosphere can be at a temperature 16% below the blackbody temperature, or 19% above it, or anywhere in between, depending on parameters. This example serves to show that the atmospheric temperature can vary widely; in our cloche model, we found that it equaled T_0, but only because of our idealized assumption of perfect absorption.

Energy-Balance Model for Planet Earth

By now, you will have got the hang of these power-budget calculations for planet-plus-atmosphere systems that are in thermal equilibrium with their surroundings.[4] In this section, we examine one that applies to our planet; it is a little more realistic than the cloche model or the thin-atmosphere model just outlined, in that it takes into account more of the features pertaining to the detailed radiation power flow that is summarized in figures 3.3 and 3.4. Yet it is still a simplification; we should accept only the broad conclusions that it leads us to and not take too seriously the quantitative predictions. Despite these weasel words, the model turns out to be quite good.

Consider figure 4.4, which depicts a simplified version of EM radiation power flow through our atmosphere. We assume that some of the incoming solar power is reflected off the atmosphere. We also assume realistic absorption coefficients. We neglect effects such as reflection off the surface, present in figures 3.3 and 3.4, and also thermals and evapotranspiration (that is, nonradiative energy transfer). Assume the mean value for incoming shortwave power density: $P_0 = 341 \text{ W m}^{-2}$. Crucially, we permit some nonuniformity of the atmosphere in that we allow the top and bottom to have different temperatures. (The bottom of the atmosphere is not meant to be the surface; rather it is the middle—in some sense—of the troposphere, where low-level clouds and air radiate longwave power down to the surface. The top of the atmosphere is here the tops of the clouds, somewhere in the upper troposphere, from which power radiates to space.)[5]

If the Earth–atmosphere system is at thermal equilibrium, then the power that flows into the atmosphere must equal the power that flows out, and the power that flows into the surface must equal the power that flows out. The math can again be found in the appendix; here are the results of this calculation. Assuming realistic values for atmospheric absorption of shortwave radiation ($a = 0.23$) and longwave radiation ($A = 0.90$), and for reflection of shortwave radiation off the atmosphere

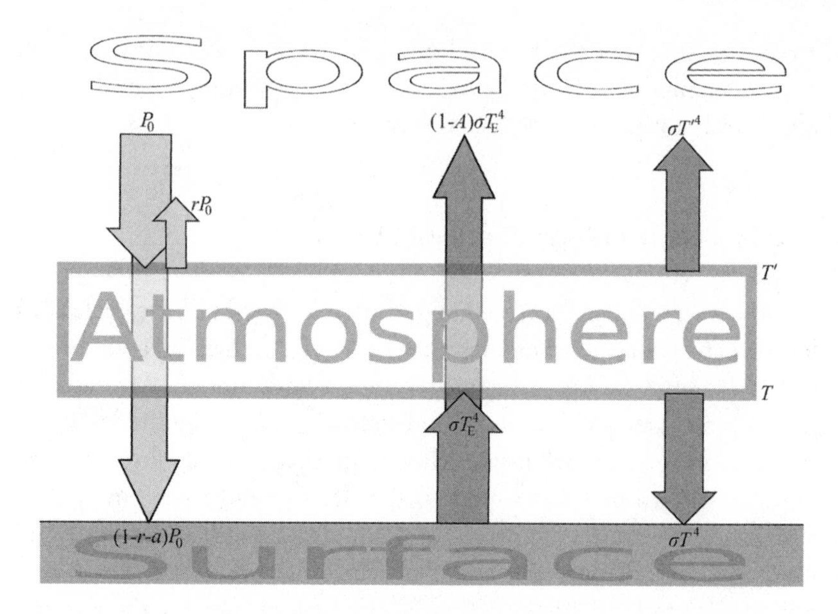

Figure 4.4 Power-balance model of the Earth–atmosphere system. Incoming shortwave radiation is reflected off the atmosphere and absorbed by it; the remainder is absorbed at the surface. Outgoing longwave radiation from the surface is partly absorbed by the atmosphere and partly passes through it, out to space. The atmosphere reradiates the absorbed power; in this model, we allow the amount radiating upward to differ from the amount heading down to the surface, by allowing for different temperatures at the top and the bottom of the atmosphere.

(r = 0.30), we find that the temperature of the lower atmosphere is T = –19°C (–2.2°F) and that of the upper atmosphere is T' = –40°C. Thus our power-balance model predicts correctly that the atmosphere should cool with increasing altitude. It also shows how surface temperature depends on atmospheric absorption and reflection parameters: T_E increases as A increases (for example, due to increased atmospheric greenhouse gases) and as r decreases (for example, due to changing global cloud cover).

If we add in the observed atmospheric environmental lapse rate (the fact that temperature falls 6.5°C [11.7°F] with every 1-kilometer [0.6-mile] increase in altitude above the surface), then we can squeeze a little more juice out of the model, or at least show that it is making sensible predictions. The lapse rate and the predicted atmospheric temperatures combine to tell us that the upper atmosphere is here at

8.6 kilometers (5.3 miles) and the lower atmosphere at a 5.4-kilometer (3.4-mile) altitude. These numbers are reasonable; 8.6 kilometers is close to the tropopause, and 5.4 kilometers is somewhere near the middle of the troposphere.

From the two atmospheric temperatures we can also see why the "back radiation" shown in figure 3.4—the longwave radiation absorbed by the surface—exceeds the radiation sent back into space. That is, in contrast to the cloche model and the thin-atmosphere model, where we *assumed* that the atmospheric radiation power transmitted upward equaled the power transmitted downward, here we predict that downward radiation is larger. From the calculation in the appendix it is 19% larger, for our model. In fact the measured value for our atmosphere is much greater than this: from figure 3.4 we see that back-radiation power exceeds the longwave power that is radiated from the atmosphere out into space by some 67%. So, the model does not capture all the nuances of Earth's atmospheric power budget—but given its simplicity we can hardly expect that it would do so.

Snowball Earth

So much for the simplest power-balance models; now we must increase the complexity to obtain a more detailed understanding and to be able to make more detailed predictions. The cost of this increased model performance is reduced transparency. The physics becomes more complex, and the equations are more numerous and more complicated. They cannot be solved with simple algebra as before; from here on, we rely increasingly on computers to number-crunch the solutions for us.

The simplest power-balance model is described as being "zero-dimensional" (0-D). It treats Earth and the atmosphere as if they were a single point characterized by just a couple of parameters (albedo, absorption). Next we added a vertical dimension by assigning different thermal characteristics to the atmosphere and the surface. Such a model could be described as one-dimensional (1-D), though this term is usually reserved for models that include many different points along the 1-D line, such as the latitudinal model we are about to investigate. The simple 1-D energy-balance models discussed so far extend into the vertical direction from Earth's surface but coarsely represent the atmosphere by only one or two points, or cells. The next logical step from 0-D models would be to divide the atmosphere into dozens of layers, assign appropriate thermal parameters to each layer, and develop a true 1-D model that covers the vertical direction in detail. However, geometry intervenes and leads us to choose a different direction.

Let us consider Earth to be divided into many different latitudinal zones. Thus we represent Earth in one dimension not by a vertical line but by a north–south line. Early climate models were greatly limited by available computer power, which is why the representation of Earth was limited to a single dimension. In the 1960s, a significant step forward in climate modeling was made when predictions for a (crude, by today's standards) 1-D north–south model were presented, which taught us about a key dynamical feature of the ice ages.

Why choose north–south as the dimension for analysis and not another dimension such as up–down? As we have seen, because of geometry the different latitudes receive different amounts of solar power. (They also have different albedos because polar latitudes are covered in ice or snow, which reflects more than rock or ocean.) This differential heating leads to current flows in the ocean and to circulation patterns in the atmosphere that transport heat from the equator to the poles. Some interesting climate consequences follow from this difference between latitudes. In a north–south climate model, Earth is divided into a number of latitudinal cells; the different cells are specified not only by different incoming solar power densities and albedos but also by the rate at which heat flows between them. Thus the number of data points necessary to specify the model increases as the size of the model (number of cells) increases. Indeed, the amount of input data that is required increases *faster* than the number of cells in the model. Thus for each cell (each latitude cell, in the present case), we need the static parameters for that cell (irradiation, albedo, absorption), but we also need parameters that describe the dynamics (heat flow between cells). Much more on this topic is in chapter 5.

In the 1960s, Soviet climatologist Mikhail Budyko investigated the effects of changing solar power output on Earth's climate and, in particular, the role that solar power variations played in generating ice ages. His 1-D model results demonstrated the complex and nonlinear nature of climate physics and brought to the fore the importance of *feedback*. Astrophysics tells us that our sun was fainter in the distant past—emitting less power—and Budyko's model showed how the average global temperature changed with changing solar power output. We are not talking about small fluctuations in solar power, such as the 0.1% sunspot cycle, but about large changes over eons of time. Thus 4 billion years ago, our sun emitted 30% less power than it does today; 2.8 billion years ago, our star's brightness had increased to 85% of its current value. In the distant future, the sun will grow bigger, brighter, and hotter still. So what did the 1-D model predict that these changes in solar power would have for Earth's average temperature? The predictions are sketched in figure 4.5.

Initially, the sun was so faint that Earth's blackbody temperature was well below the freezing point of water, and so our planet was covered in snow and ice—this is "snowball Earth." As the sun slowly warmed, Earth slowly heated up—no surprise there. But then with increasing power, the rate of heating became faster and faster, as shown by the lower curve of figure 4.5. Here is what was happening. Warmer temperatures caused ice

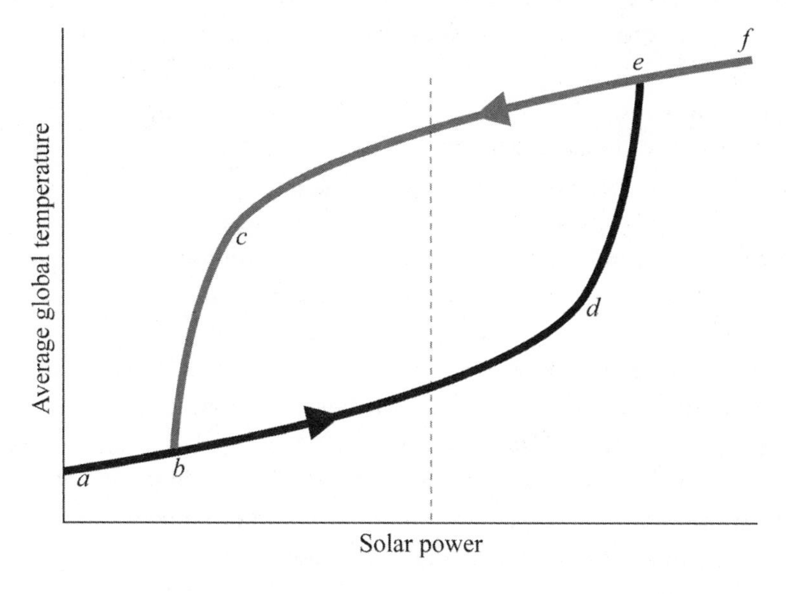

Figure 4.5 Global average temperature versus solar power output, according to Mikhail Budyko's 1-D energy balance model. (This is a sketch of the important features, not a detailed model prediction.) If the sun heats up, then global temperature follows the path *abdef*; if it cools, then the temperature does not simply reverse course, but instead follows the path *fecba*. Rapid change in ice and snow coverage, due to positive feedback, occurs at *cb* and *de*, as described in the text. The present day is indicated by the *dashed line*.

to melt, initially at the equator where solar power is greatest, and so equatorial albedo decreased. Thus heat absorption increased, and so temperatures rose faster. This is an example of positive feedback: change induces faster change. Increased temperatures led to reduced albedo, which led to increased heat absorption, which led to increased temperatures—hence the rising curve of figure 4.5. At some point, all the ice melts and so the positive feedback ends; thereafter, any further increase in solar power simply leads to a gradual increase in average temperature, as before.

Here is the surprise. Suppose the solar power reaching Earth's surface gradually decreased again, to very low values. The model predicted that average temperature would not simply reverse course but would instead follow a different path (the upper curve in figure 4.5). The reason for this behavior is the same feedback mechanism as before: falling temperatures cause ice and snow coverage to develop, beginning at the poles and then spreading toward the equator. This increase in ice and snow coverage leads to increased albedo and so to decreased power absorption and so to faster fall in temperatures—hence the accelerating fall in temperatures shown in the upper curve of figure 4.5 as solar power decreases. The rate at which temperature changes depends on the fraction of Earth's surface that is covered with ice and snow—hence the different curves when solar power is rising and when it is falling. Physicists call such behavior *hysteresis*, and it is the hallmark of nonlinear feedback; this example was an early indication of the complexity of climate physics, and of the simplicity of the Budyko model.

The consequence of hysteresis, clear from figure 4.5, is that over a wide range of solar power outputs, there are two possible values for the temperature of our planet. Which value applies at a given time depends on history: if Earth is cooling (developing ice and snow coverage), then the upper curve applies; if Earth is warming (ice and snow receding), then the lower curve applies. This behavior has important consequences, which we will unfold in stages.

All Change

The first consequence is called the *faint young sun paradox* and arose by considering the numbers that emerged from Budyko's model. Consider figure 4.5 again. The solar power difference between b and e is considerable—perhaps 30% of current solar power output. If snowball Earth really did exist in the distant past (and the geological evidence is that a frozen globe, or something very close to it, has existed at least twice—the last time 650 million years ago), then we should now be on the lower curve of figure 4.5 where it intersects the dashed line. But this point on the curve precedes the predicted period of runaway global warming (*de*) that rapidly melts the snow and ice, so we should still be a snowball. The problem is that it takes a huge boost in received solar power to melt the snowball—once in such a state, it is very difficult to get out of it. Yet here we are, in the Holocene epoch, with most of Earth ice-free. How is this possible? The increase in solar power output is not nearly enough to melt the snowball according to the Budyko model, hence the paradox.

The model is not wrong, but it is incomplete in that it does not take into account factors other than solar power output and Earth's albedo. Before listing these factors, which act to boost Earth's surface temperature beyond the energy-balance level predicted by the 1-D model, note that the problem is worse than stated: there is geological evidence for the existence of liquid water on Earth billions of years ago, when the sun was much fainter than it is today. Thus liquid water existed before the sun was able to maintain it in the liquid state, let alone melt all the ice and snow that has, from time to time, turned our planet white.

There are several factors, external to the 1-D climate model, that among them can account for the liquid water on the surface of our planet, past and present. The exact mix of these factors that actually did get us out of a snowball-Earth existence is still being debated. Take your pick from the following:

1. In the early eons of Earth's existence, the moon was much closer than it is today, as we saw in chapter 2. This fact means that tidal forces will have been stronger, generating more heat than they do today.
2. Heating due to the decay of interior radioactive elements was much stronger in the past (precisely because radioactive elements have decayed [see chapter 1]).
3. It is possible that the early atmosphere contained much more carbon dioxide than it does at present, generating a much greater greenhouse effect. (Ice covering the snowball Earth interrupts the carbon cycle, leading to an accumulation of atmospheric carbon dioxide.)

4. Before a snowball Earth arose, albedo of the surface may have been lower than at present, so heat absorption was greater.
5. Volcanic activity may have significantly heated the young Earth, though today the emissions from volcanoes act in the opposite direction.
6. Orbital perturbations may have been more severe in the early days of the solar system, leading to periods of more extreme heating and cooling.
7. The young sun may have been more temperamental—blowing hot and cold (well, less hot).
8. Plate tectonics slowly shifts the continents. In earlier times, absorption of solar power would have been greater when the continental land masses were more closely arrayed around the equator.

With one exception, I will leave these astrophysical and geological factors without further explanation; interested readers may refer to the bibliography for details.[6] The exception is volcanoes, because they have figured more than once in recent climate changes and because they illustrate an important point about climate modeling.

Volcanoes spew vast amounts of ash and gas from Earth's mantle and crust high up into the atmosphere and across the surface. This is happening now and always: the eruption shown in figure 4.6 was photographed earlier in the week in which I wrote these words. It is not an exceptional eruption, but the photograph trenchantly displays the effects. First, note that ejecta (the name volcanologists give to the outpourings) is easily above cloud level—indeed, powerful volcanoes can throw billions of tons of this matter into the troposphere and stratosphere. Second, it can be dense enough to block sunlight. Blocking sunlight is the main effect as far as climate is concerned (though not because of plume density, but because of chemical reactions in the stratosphere); it reduces the solar power reaching the surface. The immediate increase in albedo is clear from the sunlit side of the plume in figure 4.6. Eruptions are measured in hundreds of cubic kilometers and, once the ejecta from a large volcanic eruption reach the upper atmosphere, they can travel around the world, affecting global climate for several years. Thus the climactic explosion of Mount Pinatubo in the Philippines in June 1991 threw ash to a 19-kilometer (12-mile) altitude and reduced the average global temperature by 0.4°C (0.7°F) for two or three years. The eruption of a supervolcano (one that spews

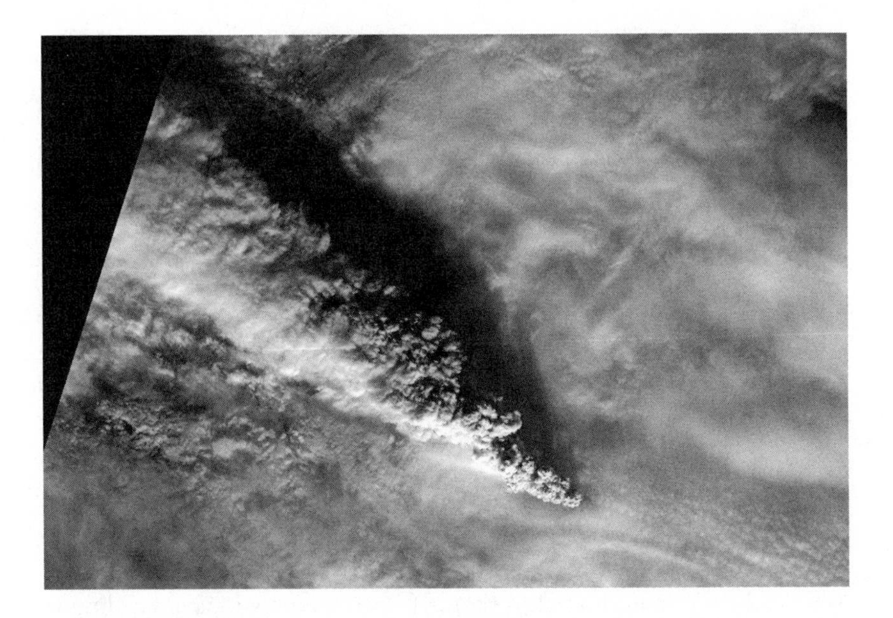

Figure 4.6 Alaska's most active volcano, Pavlof, spews ash 9 kilometers (5.5 miles) into the atmosphere on November 15, 2014. The plume is dense enough to affect air travel. (From NASA, Earth Observatory, http://earthobservatory.nasa.gov/NaturalHazards/view.php?id=84747)

more than 1,000 cubic kilometers [240 cubic miles] of ejecta), Krakatoa, in Java in 1883 threw enough ash into the upper atmosphere that it was carried by a jet stream around the world—this visible sign was an early indicator to scientists of the existence of jet streams. Snowfall levels around the world increased for five to six years following this eruption. Historical events can be linked directly to volcanic eruptions on the other side of the world. The "year without a summer" in Europe and North America, 1816, resulted in crop failures and much human hardship. It followed the eruption of Mount Tambora (Indonesia) in 1815, which lowered the average global temperature by an estimated 0.4° to 0.7°C (0.7° to 1.3°F). An eruption in Peru in 1600 likely led to climate change that caused crop failures and starvation in Russia. The strange weather reported from many cultures around the world in the year 535/536 has recently been linked to a volcanic event.[7] Geophysical records point to truly enormous prehistoric eruptions: the Lake Toba event in Sumatra, about 72,000 years ago, likely reduced global temperatures by 1°C (1.8°F) for six years.

The effects on climate of a volcanic eruption depend on the constitution of the atmosphere at the time of eruption. The gaseous emissions from volcanoes include sulfur dioxide, carbon dioxide, and hydrogen. In the very early Earth, one that should have been a snowball because of the faint young sun, volcanic activity may have significantly warmed the surface. This is because the early atmosphere consisted largely of nitrogen, with little or no oxygen. In such an atmosphere, hydrogen gas from volcanoes can react chemically with nitrogen in the air to form dimers that are strong greenhouse gases—hence the warming effect of early volcanoes (aided, perhaps, by an equatorial belt of dark volcanic ash).

In our current atmosphere, hydrogen quickly reacts with oxygen to produce water vapor. The emitted carbon dioxide is rather small (two orders of magnitude less than the human emissions, according to NASA);[8] you might think it would lead to a spike in global temperature, but such has never been detected. The sulfur dioxide, however, has a significant cooling effect. Within a month of emission, it reacts with water and sunlight to form droplets of sulfuric acid that are highly reflective, thus increasing the atmospheric albedo and so reducing the power of shortwave radiation that reaches the surface, reducing global temperatures. Sulfur dioxide in the troposphere is quickly washed out of the air, but stratospheric sulfur dioxide can stay aloft for two years or so.[9] Pound for pound, it is many orders of magnitude more effective in lowering global temperature than carbon dioxide is in raising it. For this reason, sulfur dioxide seeding of the stratosphere has been proposed as a geoengineering counter to global warming.

The influence of volcanoes on climate is significant and ever present. There are dozens of large, active volcanoes emitting gas and ash into the atmosphere as you read these words (Kilauea in Hawaii has been emitting continuously since 1983; Mount Yasur on Vanuatu has been near-continuous for 800 years; Mount Etna in Sicily has been active since antiquity). This continual seeping of gas and ash into the atmosphere is punctuated by the much more sudden eruptions of individual supervolcanoes, such as we have described. These geological events give pause for thought for climatologists who seek to predict future developments: How can we account for significant yet unpredictable occurrences such as volcanic eruptions? Hold that thought—we will come to grips with it in chapter 5 and formulate the problem more precisely in the next section, which deals with the full, 3-D modeling of the global climate.

General Circulation Models

Small Steps into the Future

General Circulation Models (GCMs) are the full-blown 3-D mathematical models of global atmospheric and oceanic circulations. They are not static energy-balance models such as we examine in the appendix, but are instead dynamic models of the physics of movement of the air and water. The Budyko 1-D model can technically be described as dynamic in that it investigated the glacial (literally) change of ice coverage, but GCMs are in a different league, if not a different world. The equations of motion that must be solved are the infamous (because difficult to solve) Navier–Stokes equations of fluid dynamics. These equations usually yield solutions only numerically—that is, when subjected to computer number-crunching rather than paper-and-pen analysis—and here they must be solved for fluids that are on a spinning sphere with irregular boundary conditions (the Earth surface and ocean floor).

So GCMs evaluate the dynamics of air and water motion in three dimensions by solving numerically the difficult equations of motion. The three dimensions are latitude, longitude, and height above the surface (or depth beneath sea level). Thus Earth is mathematically divided into 3-D cells that cover the surface and stack from the seabed up to the top of the troposphere. For each cell, the equations are solved to tell us about fluid motion between cells. As computing power has increased, so too has the number of cells that have been used in these models, with a consequential increase in the prediction accuracy of GCMs.

In the rest of this section, we unpack that last sentence a little more; there is much information in it. Numerical models of dynamical systems work as follows. The system (ocean, atmosphere, surface) is specified at some initial time—say, today at 9:00 A.M. (let us call this time t_0). That is, the state of every cell in the system is assigned an initial temperature, pressure, fluid speed, and density, and the volume and speed of fluid entering and leaving the cell are specified. In addition, boundary conditions are set (the shape, albedo, and other thermal properties of the land and the shape of the ocean floors). In modern GCMs, there may be 500,000 initial conditions that must be set to specify the atmosphere–ocean system. Now the computer crunches away and calculates the state of the system at some time t_1 in the near future, where $t_1 = t_0 + \Delta t$, where Δt is a small *timestep*. The choice of Δt is limited by cell size: smaller cell size requires smaller Δt. This is because, to capture the physics correctly, it is necessary that the calculations take into account the movement of material from one cell to an adjacent cell. So if a polar jet stream can move at a speed of 250 kilometers (155 miles) per hour and the cell size is 250 kilometers, then the time step cannot be larger than 1 hour.

At this stage, our GCM math model has calculated the new value of variables for all the cells at a time that has advanced by one time step from the initial time. Now the process is repeated again and again so that the state of the atmosphere–ocean system is predicted for some significant interval into the future—say, for 24 hours in the case of weather forecasting or 10 years in the case of climate modeling. Thus our computer model is able to predict future climate (or weather) based on current conditions plus our knowledge of the physics of how this complex system evolves.

Numerical Weather Prediction

Some of the same equations are used for both numerical weather prediction (NWP) models and GCMs; sometimes even the same computer code is employed in both. This chapter is mostly about climate modeling, but it is appropriate to devote a few lines here to weather modeling because the numerical processes are so similar.

The differences between the two types of modeling are due to the difference in timescales. Thus weather predictions are said to be short scale (1 to 3 days) or medium scale (4 to 10 days), but these are minuscule compared with climate timescales, which are measured in decades or centuries. Thus modeling of ocean dynamics is not necessary for weather prediction because the ocean does not change much over short intervals; instead, it is sufficient to specify the ocean temperatures for each surface cell of the atmosphere—what goes on underneath is not of interest. Another difference between NWP and GCMs is that accurate knowledge of initial conditions is crucial for weather prediction but not for climate predictions. Why is this so? For climate, errors in the initial conditions often fade away as long as they are not too big. Climate is average weather, you may recall, and so detailed knowledge of transient events is not needed. Weather *is* transient events, so if you start out with the wrong state then, a few hours later, you will still be in the wrong state.

Cell sizes are smaller for weather prediction—say, 40 kilometers (25 miles) for a modern NWP model as opposed to 200 kilometers (125 miles) for a GCM. This increased resolution is required because meteorological phenomena are often on smaller length scales; a thunderstorm is of great interest to a weather forecaster but is too small and brief for a climate modeler to bother about.[10] Timesteps are smaller because cell sizes are smaller. Also, it is necessary for weather models to run in only a few hours, because predictions are for only a few days into the future—there is no point in trying to predict tomorrow's weather if the program takes a week to run.

Evolution of Modeling

Climate modeling began in the late 1960s and has evolved greatly since that time, in part because of an increased understanding of the physics that underpins climate and climate change. The evolution can, however, be attributed mostly to the rapid increase in computer power over the decades. In fact, computer speeds have increased by a factor of 100 every decade since 1950. Moore's Law has operated quite accurately since 1970; this law is really an empirical observation about the density of transistors that can be placed on a microchip of given size, but it can also be interpreted as the rate of growth of computer memory. According to this interpretation, computer memory approximately doubles every two years. Thus computers today are much faster and have vastly greater memories than did those in past decades. Consequently, the number of cells that can be included in a GCM has increased, and so the size of each cell has become smaller over the decades.[11]

Resolution cells are not necessarily cubes, or rectangular boxes, though it is easiest to picture them in this way. They can have different shapes, and their sizes can vary within a model. Cell sizes in the lower atmosphere may be different from those in the upper atmosphere (the vertical extent usually differs); cell size near the surface of oceans may vary with distance from coastlines. (For weather forecasting, cell sizes near population centers may be smaller than those over unpopulated land or open ocean.) A good cell size for a GCM today (2015) is about 1° latitude or longitude over land (more over water), perhaps 40 to 60 layers in the atmosphere, and 30 layers in the ocean.

The accuracy of GCM predictions has improved as a direct consequence of cell-size reduction, but also indirectly. The earliest computer models in the 1960s had coarse horizontal resolution and only two atmospheric layers. Their predictions missed many features of real climate systems but correctly reconstructed basic wind patterns. The manner in which water vapor is conveyed to the upper atmosphere was first modeled during this decade. By the 1970s, some models included nine layers for the atmosphere plus some crude geographical surface features on the scale of continental landmasses. The extent of sea ice became calculable. One GCM from this period ran for 50 days to produce a 300-year climate forecast. In the early 1980s, the oceans were still not modeled in detail. Instead, they were considered to be a slab with a surface temperature—no attempt was made to determine the flow of currents and heat that happened at depth. Today ocean models have up to 30 layers with cell sizes that are smaller than those of the atmosphere (the horizontal area covered by one atmospheric cell requires as many as six ocean cells). These ocean models connect to atmospheric models so that, for example, heat flow from the ocean to the lower atmosphere is modeled properly rather than being input as external parameters, as before.

Since the 1980s, gravity waves (large-scale ocean waves, buoyancy waves in the atmosphere) have been included in models—their importance is now realized. More and more geophysical phenomena are being brought into GCMs such as the carbon cycle and sea-ice changes. Previously, the influence of these phenomena was *parameterized*, meaning that relevant empirical data pertaining to them were fed into the model as external influences; now they are properly modeled and so can be included in the predictions of the model. From the 1990s, soil and vegetation models have entered GCMs and have resulted in more accurate predictions. Evapotranspiration—the contribution of photosynthesizing life-forms to atmospheric carbon dioxide and water vapor—is being incorporated. The El Niño phenomenon is being included in coupled atmosphere–ocean GCMs. As these "extra" phenomena become fully integrated into the models, the abbreviation GCM evolves from General Circulation Model to Global Climate Model.[12]

Accuracy of Modeling

The accuracy of GCMs and of their weather equivalents can be questioned. How can these models be tested? Let us say that the accuracy of a GCM is estimated by how closely (in some statistical sense) its predictions match the actual state of the climate as measured by detailed observations at some particular time in the future. It is reasonable to suppose that the accuracy of the predictions will become lower and lower as the period of time increases. That is, our GCM may predict the state of our global climate to within 20% of measured values in 10 years, but the divergence from actuality will increase at 20 years, and increase more at 100 years. We will delve into numerical considerations of weather and climate prediction accuracy in chapter 6; here we summarize qualitatively the accuracy of GCM predictions and raise issues to be addressed later.

We have already encountered a serious issue for the accuracy of model predictions in the simple 1-D Budyko model of ice-age evolution. Recall that this model raised a question about how an ice age, once developed, could end, leading to the question of how Earth could have been warm enough to include liquid water on the surface back in the days of the faint young sun. One possible explanation was the eruption of volcanoes. But what does this awkward fact of climate development do for GCM predictability? Our GCM required vast amounts of initial data to set it going; now we are saying it needs more than this. The hysteresis characteristic of climate dynamics means that historical events shape current climate evolution—we need to know data from past ages as well as current data. Furthermore, we cannot predict further into the future than the next supervolcano eruption. We may develop a really accurate climate model that can tell us the climate in 50 years to within 10%, but that accuracy goes out the window in the event of the Great Yellowstone Eruption of 2040. We do not have, and will not have in the foreseeable future, a GCM for the core and mantle of Earth's interior telling us when volcanoes (and earthquakes) will occur. Hence volcanic perturbations to our climate must always be parameterized—added in as external happenings based on observations.

One way of estimating the accuracy of GCMs is by *hindcasting*: choose the start date t_0 for a model run to be sometime in the past, say January 1, 1800. Then we run the model forward, including unforeseeable events such as the Tambora eruption of 1815 and the Krakatoa eruption of 1883. We can compare predicted climate developments with detailed measurements of the climate for the years between 1800 and the present day. Once satisfied with the accuracy of our GCM, we can then play with the model, introducing historical "what-ifs" to see, for example, the state of the climate today if humans never developed the industrial revolution, or if our industries emitted 10 times as much carbon dioxide since 1950, or if an asteroid struck Earth on Millennium Day, 2000. We can cautiously accept predictions of our GCMs for the future (barring unforeseeable events, such as the Great Yellowstone Eruption of 2040), as well as accept them for these "what-if" scenarios, if and only if our models can reproduce very well the known historical climate trends.

So how well do our current GCMs perform? They are very good at predicting tropospheric air temperatures; they are good at predicting regional air pressures and precipitation rates (within about 25%). The averaged prediction of many different GCMs is generally better than the predictions of individual GCMs. All models have shortcomings in predicting the stratospheric climate.[13] The jet streams are predicted but not very well (their shape and separation are wide of the mark). A major issue with current models is with the treatment of clouds; inaccuracies in this treatment lead to inaccuracies in model predictions. Clouds (investigated in chapter 7) are the result of convection aloft of warm, moist air. The trouble is that convection events are too small-scale to be modeled at all by GCMs. (This is a problem of resolution—of climate model cell sizes: NWPs can foresee convective thunderstorms, after all.) Thus cloud cover is parameterized; this parameterization is more sophisticated than simply assuming average levels of clouds with typical albedos and latent heat content, but it is still just a parameterization. Clouds are a particular problem because they lead to competing effects (they increase albedo, which lowers atmospheric heat input, and they increase back radiation, which raises it) and so it is important to include them correctly. Ideally, and hopefully one day in practice, resolution will be good enough for the models to predict clouds arising and disappearing so that inserting cloud parameters will be unnecessary and the accompanying inaccuracies will never arise.

Past and Future Climate Change

The current climate debate has been ongoing for four decades at least; it is widespread and polarized. There are two threads. First, how accurate and consistent are the data, and what do they tell us about the current climate? Does global warming exist and is it significant? Second, if global warming exists, how much of it is anthropogenic—due to human activity? What do the GCMs tell us about future climate based on the data we have for current climate, and can we believe these predictions?

Let us deal with the data question first. We can observe current weather conditions and maintain records of these observations; this has been going on with increasing sophistication and detail for a couple of centuries. Average these records over, say, a 30-year sliding window[14] and we have climate data. There are no (or insufficient) weather records for earlier time periods, and so climate has to be inferred from other types of data, known as *proxy data*. Thus we can infer certain climate details, such as precipitation rates from tree-ring data, back from the present day to about 5,000 years ago. Other information such as oceanic carbon dioxide concentration comes from studying sediments from seabeds that contain coral and plankton skeletons. This source provides information that goes back about 20,000 years. Air trapped in glaciers or continental ice sheets provides direct samples of the atmosphere going back as far as 160,000 years. Deep-sea sediments provide more proxy data that can tell us something about the climate as long ago as 170 million years. Fossils yield useful climate information back as far as 550 million years.[15]

Couple all these data with geological evidence (for, say, the distribution of continents and of different minerals) and astrophysical data (such as the orbital wobbles discussed in chapter 1) and it is not hard to see how climatologists can be fairly confident about the general features of Earth's climate in past eras—and very confident about the climate in recent centuries.

So what do the data—direct and proxy—tell us? Earth's climate is in flux and always has been. The variability operates on many timescales, some very slow (eons) and some very fast (a few years). We have seen that the distant past has exhibited snowball-Earth climate conditions; at the other extreme there have been several periods (such as 145 million years ago, and 50 to 40 million years ago) when there has been no polar ice: Greenland and Antarctica were ice-free. The ice-free periods obviously correspond to warm climates, with sea levels much higher than today because water expands when warm but also because of the extra water freed from melting polar ice caps.

The last glacial maximum occurred 21,000 years ago, when the average global temperature was 3° to 5°C (5.4° to 9°F) cooler than it is today. Since then, the climate slowly warmed until about 10,000 years ago, when the temperature reached 15°C (59°F). We have been in this interglacial warm period from that time; a running average of the global temperature has wobbled about the 15°C level, but always within 1.5 degrees. The warming of the world following the last glacial maximum is plotted in figure 4.7, based on the average of nine proxy-data reconstructions.

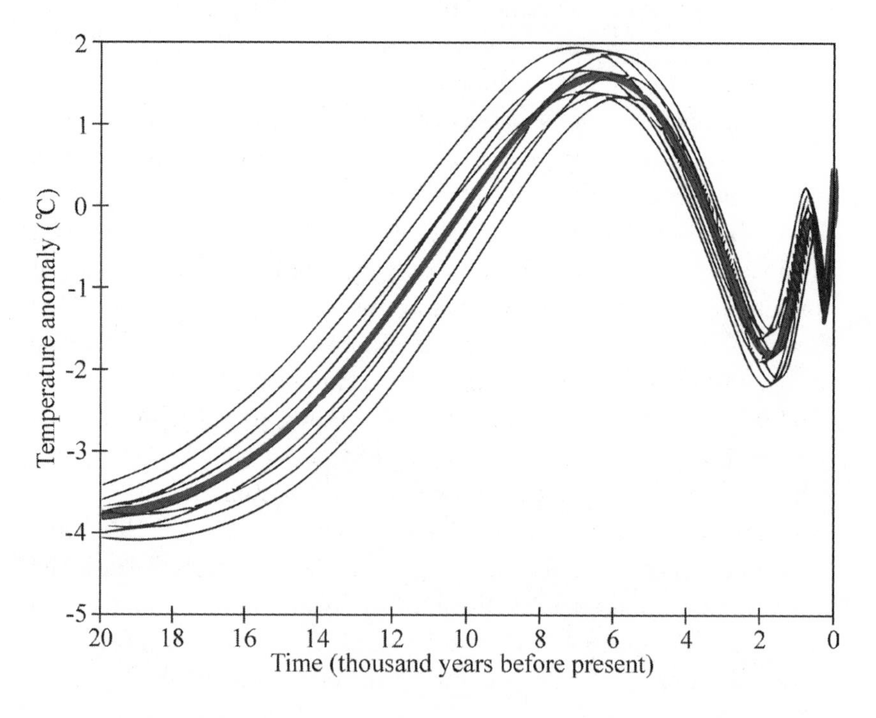

Figure 4.7 Global average temperature (*thick line*) versus time, over the past 20,000 years. The temperature is referenced to the average for the years 1961 to 1990. Temperatures are reconstructed from nine different proxies (*thin lines*), taken from the National Oceanic and Atmospheric Administration climate database. (Adapted from S. P. Huang et al., "A Late Quaternary Climate Reconstruction Based on Borehole Heat Flux Data, Borehole Temperature Data, and the Instrumental Record," *Geophysical Research Letters* 35 [2008]: L13703)

The Medieval Warm Period occurred about 1,000 years ago, during which time northern Europe, including Iceland and Greenland, was warmer than it was centuries earlier and later (though not as warm as it is today). Conversely, temperatures were lower during the Little Ice Age of the seventeenth and eighteenth centuries. The temperature changes of the past 400 years are shown in figure 4.8 as estimated from proxy data of glacial extent. Statistically, the most significant feature of the temperature curve is the warming that has occurred in the twentieth century (a feature that is reproduced in many other proxy-data reconstructions of global temperature). This abrupt rise means that the current average temperature, at least in the Northern Hemisphere, has been the highest in the past 500 years and perhaps in the past 1,300 years.

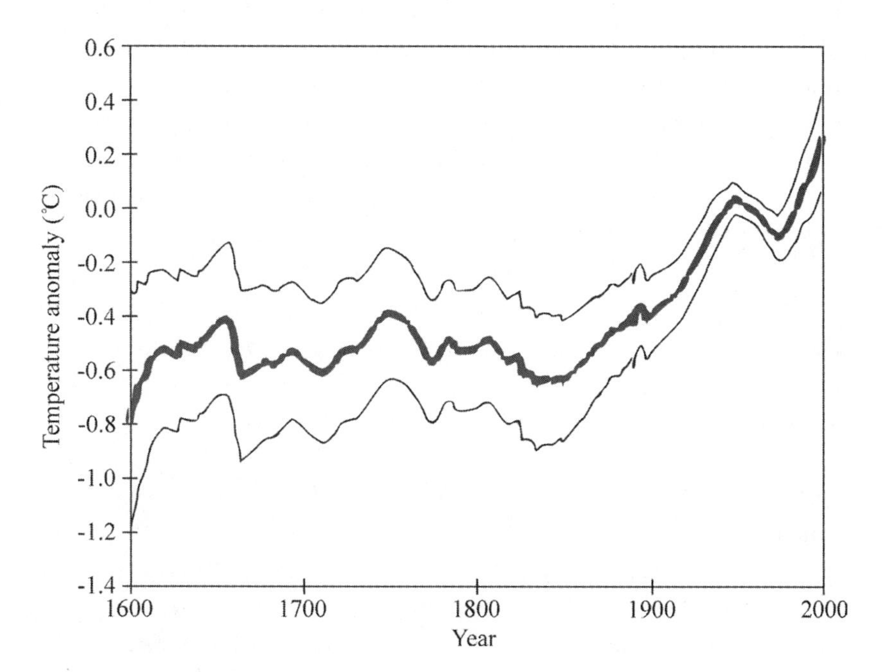

Figure 4.8 Global average temperature versus time (*bold line*), over the past 400 years. The temperature is referenced to the average for the years 1961 to 1990. The *thin lines* denote estimation errors (95% confidence). Temperatures are reconstructed from glacier length variation proxy, taken from the National Oceanic and Atmospheric Administration climate database. (Adapted from P. W. Leclercq and J. Oerlemans, "Global and Hemispheric Temperature Reconstruction from Glacier Length Fluctuations," *Climate Dynamics* 38 [2012]: 1065–1079)

The paleoclimate record of the distant past displays other abrupt temperature changes; almost all climatologists attribute the current rise mainly to human activity.

And so we come to the second thread of the climate debate. Why do climatologists (as reported by the Intergovernmental Panel on Climate Change [IPCC], endorsed by the American Meteorological Society, the American Geophysical Union, and the American Association for the Advancement of Science) now state that the current abrupt temperature rise is due to human activity? In a nutshell, here is the method. First they apply GCMs with a start date before the onset of the first industrial revolution (say, 1750) and produce predictions of global average temperature from this start date to the present day that are close to observations; this validates the models. Second, they repeat the model runs except that they remove contributions to the atmosphere that originate from human activity, such as the carbon dioxide and other greenhouse gases that industry has generated over the past couple of centuries. The result is a temperature curve that departs increasingly from the actual historical data from about 1950. Without the anthropogenic greenhouse gas contribution, global average temperature today would be about 0.8°C (1.4°F) lower than it is.[16]

Thus data show that global temperature has increased in recent decades and that atmospheric greenhouse gas concentrations have increased. GCM predictions link the two, demonstrating that the extra greenhouse gases are responsible for the rise in temperature.[17]

Here I have presented the detailed argument. At a broader, higher level, climatologists observe that more heat is entering Earth's atmosphere than is leaving it and note the human contribution to atmospheric GHG levels that trap heat. Observed climate changes are those to be expected from such a "stoked up" system. The pattern is changing in ways that we can predict from climate models, but even without those models most climatologists can see that altered planetary cycles are due to excess heat. The cycles themselves (such as El Niño) move the heat around, but are not the cause of increased heat in the system—that cause is increased greenhouse gas levels.

A few skeptics remain, about the human contribution to global warming.[18] As stated in the introduction, I do not propose to debate the issue. Let me simply quote from an editorial in the international science journal *Nature*: "Many climate skeptics seem to review scientific data and studies not as scientists but as attorneys, magnifying doubts and treating incomplete explanations as falsehoods rather than signs of progress towards the truth."[19]

The Future

The current rise in global temperature has been accompanied by a rise in average sea level (by about 20 centimeters [8 inches]) and by a reduction in polar ice, as you might expect. Atmospheric carbon dioxide concentration hovered near 280 parts per million (ppm) from year 1 to 1750; since then, it has risen to 400 ppm. The figures for other greenhouse gases show a similar sharp increase, resulting in what is sometimes called a *hockey stick* graph: methane levels have risen sharply from 700 parts per billion (ppb) to 1,200 ppb, and nitrous oxide levels from 650 ppb to 1,900 ppb. GCMs predict that if carbon dioxide levels double in the future (they are expected to do so by 2100), global average temperature will increase by about 3°C (5.4°F). The increase will be greatest in polar regions.

An interesting hiatus in the temperature rise of recent decades has now been explained. For the past 15 years or so, the global temperature appeared to be more or less constant and not rising. GCMs show that this phenomenon was not due to any reduction in the amount of excess power received by Earth—that is, the difference between incoming shortwave power density and outgoing longwave power density. This excess has remained constant at about 0.8 W m^{-2} (which is about half the radiative forcing caused by human activity). It seems that some of the extra power is causing deep ocean water to heat up instead of directly heating up the atmosphere. The rate of atmospheric heating may have slowed temporarily, but new data and more refined statistical analyses have shown that the hiatus never really existed, and that the slowing will very likely not persist for much longer. What makes this "hiatus" interesting—at least to observers of human nature—is the manner in which it has been seized on by both sides of the climate change debate.[20]

The weather consequences of a further increase in global temperature include more frequent heat waves, more frequent record high temperatures, plus increased humidity and precipitation. This increase will be unevenly distributed. Generally, we can expect more weather in a warmer world, because there is more energy in the system. The temperature rise is less than it might have been because of another human influence: aerosols added to the atmosphere by human industry have the same cooling effect as volcanic aerosols. Ironically, the move to cleaner fuels is reducing this anthropogenic aerosol contribution and so is reducing the mitigation of global warming by greenhouse gases; in other words, cleaner fuels that produce greenhouse gases contribute more to global warming than do dirtier fuels, for the same amount of greenhouse gases produced.

The anticipated effects of future global warming on humans are as follows:

- Reduced freshwater resources from melting snow and ice (eventually affecting one-sixth of the world's population)
- Increased areas of drought
- Increases in extreme precipitation and flooding
- Increased number of wildfires and of building fires
- Melting permafrost, which causes pipelines and pumping stations to sink and releases greenhouse gases to the atmosphere (another positive-feedback loop)[21]
- More uncertain infrastructure logistics (for example, planning for long-term projects such as dam construction) because of changes in surface-water abundance and flow
- Initial increase in cereal crop production at middle and high latitudes
- Change in threats from diseases
- Sea-level rises, leading to increased salination of groundwater and estuaries
- Warmer and more acidic oceans, leading to higher rate of species extinctions (such as corals)

These effects are not direct consequences of the GCMs but follow logically from their predictions about changing climate and weather phenomena.[22]

***

The simplest climate models amount to energy- or power-balancing of the atmosphere and surface of Earth, analogously to greenhouse thermodynamics. Such models can explain why the lower atmosphere is warmer than the upper atmosphere and why back radiation exceeds atmospheric radiation to space. An early dynamic climate model, Budyko's 1-D ice-age model, displayed hysteresis and exposed the faint-young-sun paradox. These issues emphasize that climate models require good input data, not just for the present but also from the past (hysteresis) and for the future (volcanoes). GCMs are increasingly detailed 3-D climate models that are making increasingly accurate predictions. Current climate change is centered on rising global mean temperature and is caused largely by human activity. These changes will have consequences for future generations.

5

Oceans of Data

A small sample, we repeat, is rarely the big scientific problem. Interpretation is.
Stephen Ziliak

No scientific subject relies more on data for its predictions than climate modeling and meteorology. Here we recall why this is the case and examine how the data are gathered and applied, and how technology influences data gathering and the prediction accuracy of models. We start at the bottom—looking at the types of data that are so assiduously garnered—and work our way up to see how these data are organized and shared.

Development of Data Collection

Weather has been recorded for centuries, but the systematic measurement of meteorological phenomena began only in the nineteenth century with the establishment of government-funded observational weather networks. Rudimentary forecasts appeared by the mid-nineteenth century, mostly to provide storm warnings for mariners. New and improved meteorological instruments introduced in the early twentieth century (thermometers, barometers, hygrometers, and anemometers) heralded more precise observations of atmospheric conditions near the surface of Earth. Balloons were first employed to measure temperature and humidity at higher altitudes. Observation stations were connected by telegraph networks. Today, the suite of measuring instruments is greater, their measurement precision has improved, and the measurements are performed faster and more frequently and are reported automatically via telemetry.

Data are not of uniform significance. Thus some types of data need updating more often than others; some are more local than others. For example, wind speed and direction are recorded; this information is assumed to hold for the surrounding area out to 2 or 3 kilometers (1.2 or 1.8 miles, but a shorter distance up into the air) and for a duration of two minutes. After two minutes has elapsed, or for a point on the surface that is more than 2 or 3 kilometers distant, these data do not apply. The validity of different types of meteorological data, in space and time, is illustrated in figure 5.1.

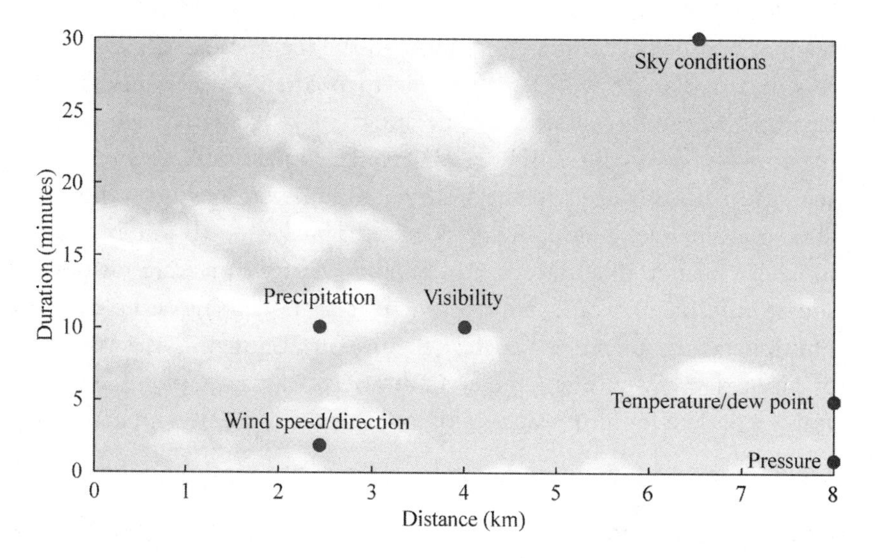

Figure 5.1 Weather data validity: duration and distance. A measurement of visibility is assumed to hold for 10 minutes at the measurement location and for locations within 4 kilometers (2.5 miles). If a measurement is made at a neighboring site—say, 5 kilometers (3 miles) distant and 15 minutes later—then the visibility at intermediate locations and times can be found by interpolating (by forming a weighted average of the two measurements). Other data are valid for shorter or longer duration, as shown, and for greater or lesser distances.

It is obvious that the quantity and quality of input data influence a numerical weather prediction: bad data in, bad prediction out. But the converse is also true in that the nature of what is being predicted influences the quality and quantity of data required. Thus, for example, droughts follow from long, persistent dry periods: daily rainfall data for a few weeks will lead to this easy prediction. But thunderstorms develop much more quickly, and so their prediction requires input data that are updated much more frequently—every few minutes. The predictability of meteorological events depends on the events, and so the quantity of data that is required for a prediction of given confidence ("there is a 90% probability of 10 millimeters [0.4 inch] of rain in your area tomorrow") depends on what is being predicted. Something that develops rapidly over a wide area requires much more input data than does something local that evolves slowly.

Generally, more and more accurate data means better predictions, and for this reason (among others) weather predictions have improved, but there is a fundamental limit to this progress. In chapter 6, we will see that weather dynamics are chaotic, in the technical sense, and why this means that we will *never* be able to predict weather months in advance. Consequently, there is a limit to how accurately we should measure our data—measuring rainfall to the nearest millimeter makes sense, but measuring it to the nearest micrometer does not. There is also, in principle, a limit to how much data we need—air pressure for every hectare of surface would be good, but air pressure for every square millimeter would be too much of a good thing. One of the limiting factors of meteorological prediction accuracy is the volume of input data. We are not there yet, however. For example, there is a scarcity of data for air pressures aloft and for open-ocean temperatures.

Land Surface Data

We will work our way up from the bottom in another sense: gathering data first from Earth's surface and then up into the atmosphere and space. Let's start with meteorological data gathering over land. If you are of a certain age, you may recall school projects involving a little weather station where you went to record manually air pressure, temperature, and rainfall over the previous 24 hours. In general, such manual data gathering is becoming a thing of the past—today's weather stations are automated or are hybrid. In the United States, the automated weather station network is known as Automatic Surface Observation System (ASOS); it was developed by the National Weather Service, the Federal Aviation Authority, and the Department of Defense. ASOS is the nation's primary surface weather observation network. Stations are installed at 900 airports around the country. Each reports up to 12 times per hour 24/7/365 on a unique frequency the following data: date and time, local wind speed and direction, visibility, sky conditions (cloud cover and height to 3,700 meters [12,000 feet]), air temperature and wind chill factor, dew point and relative humidity, air pressure, and precipitation rate.

Such automated weather stations are more consistent than human observers, which matters particularly when the weather is changing quickly.[1] The reported data are subjected to elementary statistical treatment before being recorded, so that, for example, more recent measurements are weighted more heavily than older ones. Some types of data are considered to be more important than others. Thus near an airport, the visibility is updated more often than other data because of its significance for airplane safety.

The surface weather data are integrated into a nation's weather centers for further analysis, or are sent to commercial weather services such as Weather Underground to provide real-time local weather information over the Internet for most urban centers across the world. We will see later in the chapter how local data are integrated to form an update of global weather conditions; these data are input to numerical weather prediction programs and are displayed as weather maps on your local weather channel to show weather trends and large-scale effects. Thus a single weather station will likely not spot a weather front, but the reports of many weather stations are integrated and the displayed data show weather fronts clearly.

The other main method of gathering weather data from the land surface of our planet is by means of weather radars. These remote sensors gather different types of data than the ASOS network. Radar data are generally from higher up in the atmosphere and are a near-instantaneous gathering of data from a wide swath of sky.

Radar developed as a significant aspect of military technology during World War II.[2] Some of the microwave wavelengths that were transmitted by radars during that conflict detected weather features (rain, in particular) as well as enemy aircraft. Radar works by monitoring its own electromagnetic transmissions that are reflected back to its antennas; signal-processing techniques then clean up the images for plotting on display screens. A significant challenge in radar development is the problem of *clutter*. The transmitted radar signal will reflect off the intended *target* if it is within the radar beam (figure 5.2*a*), but the transmitted signal will also reflect off a lot of unwanted environmental stuff such as mountains or clouds or rain. One person's clutter (the reflections that she doesn't want to see) is another person's target (those she does want to see). After the war, many surplus military radars were reemployed to detect weather phenomena such as precipitation.

Modern weather radars are specially designed for the purpose—they are not simply second-hand military technology that just happens to be able to detect rain. They are the most effective tool in the meteorologist's arsenal for detecting precipitation. The transmitter wavelengths (usually around 10 centimeters [4 inches]) are carefully chosen to maximize the signals received from weather features. The radar beam is narrow, and the antenna rotates so that the beam covers a large area around the radar site, up 200 kilometers (125 miles) out. In the 1990s, 150 NEXRAD (*NEX*t generation weather *RAD*ar) Doppler radars were installed in the United States, each transmitting an average power of 450 kilowatts. The data that each provides are updated every five minutes.[3] Many nations have their own weather radar networks, and if these countries have common borders then the radar data are pooled to supply a common weather radar map of the whole area—the movement of rain clouds does not respect international borders.

Doppler radars can detect the speed of the target based on the shift in frequency of the transmitter signal that this speed induces. More precisely, these radars detect the radial component of speed—that is, the component along the line of sight from radar to target. The radar cannot detect speed that is transverse to the beam, and so an object that passes across the radar beam will appear to be stationary. In fact, special software picks out storms at an early stage of their development by matching the observed signal pattern to a template. Then the storm feature is tracked so as to provide useful information about the storm speed

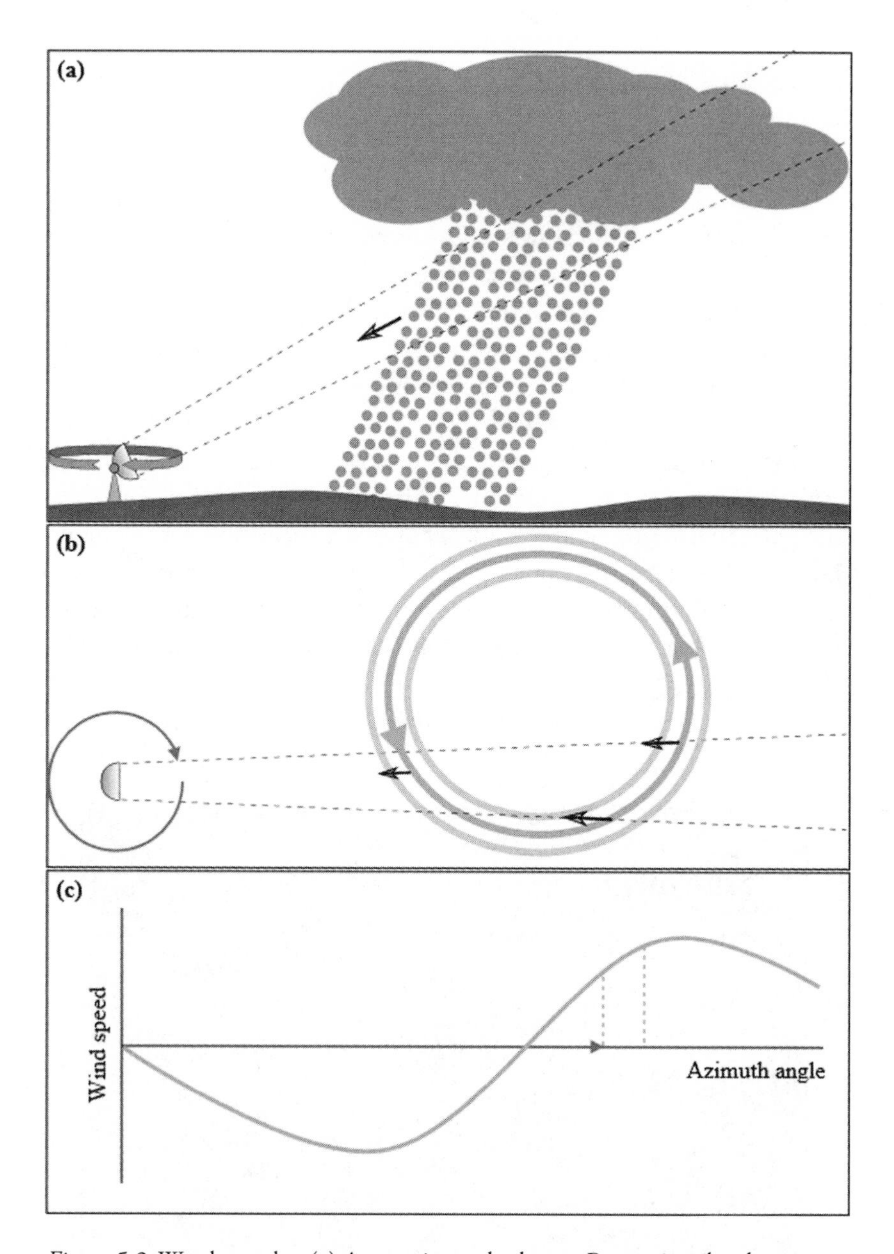

Figure 5.2 Weather radar. (*a*) A scanning radar beam. By varying the elevation angle from one scan to the next, the beam can cover the whole sky. This beam is detecting rain and can measure the rain altitude and distance, and the rainfall rate. (*b*) A scanning radar beam and a cyclonic storm, viewed from above. The Doppler radar reflections (*black arrows*) contain information about the radial component of wind speed. (*c*) The way in which wind speed varies with scanning angle for the cyclonic storm of (*b*).

and likely path, so meteorologists can predict the likely time of arrival at a particular location—say Miami, bracing itself for a hurricane.

Radar data are incorporated into numerical weather-prediction models along with data from other sources. The data from different radar locations are also displayed as weather maps, dear to television weather forecasters. These maps provide visible images of large-scale storm coverage and of precipitation extent and movement, in near real time (within a few hours of the data being gathered). An example of a national radar weather map is shown in figure 5.3.

Modern weather radars can detect more than precipitation. They can locate *virga* (rain and snow that does not reach the ground) and updraft zones (a feature of cyclonic storms). They can detect wind shear and microbursts; these are wind phenomena that appear in thunderstorms. Modern radars can thus discern the three-dimensional structure of thunderstorms and of tornadoes. They can distinguish between rotating tornadoes and straight-line winds (*derechos*), and between rain and hail. By adopting polarized transmission signals, weather radars can even detect the shape of individual rain droplets; this capability

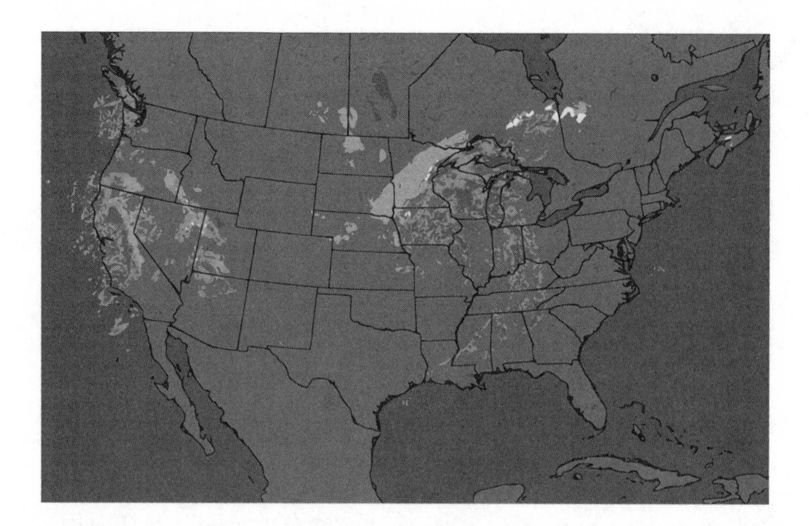

Figure 5.3 Precipitation pattern for the contiguous United States at 1:00 A.M., December 16, 2014, Coordinated Universal Time. This map is formed from the combined images of many weather-radar data sets. (From National Weather Service, Precipitation Map, National Oceanic and Atmospheric Administration, http://www.srh.noaa.gov/ridge2/RFC_Precip/)

helps to improve the estimation of precipitation rate (because drop shape depends on the fall rate) and of the mix of snow and rain or hail that is generated by a winter storm.[4]

Ocean Surface Data

The most common wind direction for the Pacific regions of the United States and Canada is westerly. Thus most of the weather for these regions comes from the Pacific Ocean. Similarly, most of the weather for western Europe comes from the North Atlantic. Obviously, this fact creates a problem: With no land in the areas where weather is formed, how do meteorologists get the raw data they need to make forecasts? Some weather data come in from ships, other data from weather stations aboard offshore oil rigs, but most are transmitted from moored buoys that are positioned strategically to obtain weather information from locations that are deemed important. The distribution of weather buoys around the United Kingdom and Ireland illustrates this point (figure 5.4).

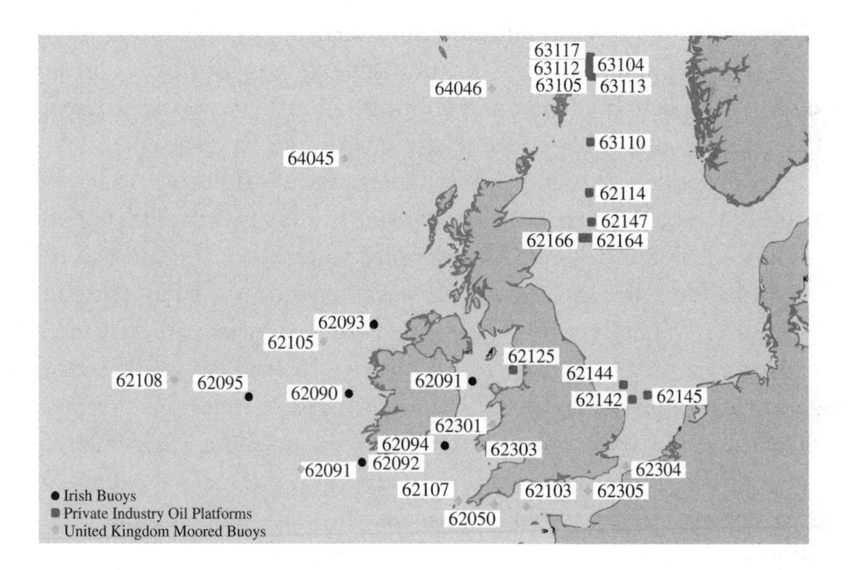

Figure 5.4 Locations of numbered British and Irish weather buoys. The data that are reported from any one of these numbered buoys, or from any others around the world, can be obtained from the National Data Buoy Center, National Oceanic and Atmospheric Administration, http://www.ndbc.noaa.gov/.

The data that a typical weather buoy measures automatically (and transmits back to meteorological centers) are current location and time, wind speed and direction, wave height and period (wave period is the interval between peaks passing the buoy), air pressure and trend (increasing or decreasing), air temperature, and water temperature.

Atmospheric Data

Moving up in the world, literally, we come to weather balloons and the radiosonde instrumentation that they carry aloft. Knowledge of the state of the troposphere is essential to numerical weather prediction and so a concerted worldwide effort to make the necessary measurements is carried out twice daily, at midnight and midday Coordinated Universal Time. At these times, 800 balloons from 1,500 or so sites around the globe are released (figure 5.5). They are made of latex and filled with helium (usually, though sometimes hydrogen) and so are lighter than air. They rise due to buoyant force and carry with them a 250-gram (0.5-pound) battery-powered radiosonde that measures air pressure and temperature as well as relative humidity. (Today, balloon latitude, longitude, and altitude data are provided by GPS tracking.) As the balloon ascends, this information is relayed by telemetry to a ground base, where it is integrated with other meteorological data. The radiosonde is suspended by cable as much as 60 meters (200 feet) beneath the balloon, so that the radiosonde measurements are not contaminated by the balloon; for example, power from the sun can heat up the balloon. Because the air thins with altitude, the balloon expands as it rises. Depending on balloon size (they start of with a diameter between 1.5 meters [5 feet] and 8 meters [26 feet]), it bursts when it gets to an altitude of 20 to 40 kilometers (12 to 25 miles). At this point, the radiosonde descends to Earth on a small parachute. Currently, data are not gathered during the descent phase. If it is recovered (some 20% are recovered in the United States), then the radiosonde may be reconditioned and reused.

If there is a crosswind, the balloon may drift several hundred kilometers downwind during its ascent. Sometimes no radiosonde is attached so as to measure wind speed and direction more accurately as

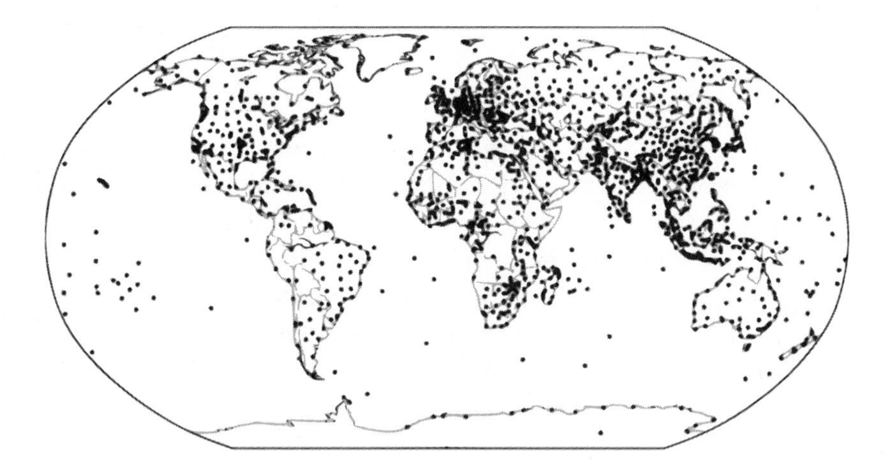

Figure 5.5 Locations of the 1,500 weather-balloon stations around the world. Among them, they carry some 800 radiosondes aloft, twice daily, at midnight and midday Coordinated Universal Time. (From National Centers for Environmental Education, "Integrated Global Radiosonde Archive," National Oceanic and Atmospheric Administration, https://www.ncdc.noaa.gov/data-access/weather-balloon/integrated-global-radiosonde-archive)

a function of altitude. In this case the balloon is known as a pilot balloon and is tracked by radar or GPS.

The radiosonde instrumentation is calibrated assuming that the balloon ascent rate is between 275 and 350 meters (900 and 1,150 feet) per minute.[5] At this rate, the balloon takes about two hours to reach an altitude of 30 to 35 kilometers (19 to 22 miles), which is typical for a balloon with a diameter of 2 meters (6.6 feet).

The United States alone releases about 70,000 weather balloons every year, from 92 stations in the continental United States and in Hawaii, American Samoa, Guam, and Puerto Rico, and also supports a further 10 stations in the Caribbean. These weather stations and all the others around the globe release their balloons simultaneously, so that the world's meteorologists can get a snapshot of the atmosphere, twice daily. This synchronization has been in effect since 1957. The sites are known as Integrated Global Radiosonde Archive (IGRA) stations, which tells us something of the cooperation between national weather agencies that has been established to obtain atmospheric meteorological data.[6]

Data from Space

Doppler radars look up at the atmosphere, probing it at microwave frequencies; weather satellites look down on the atmosphere and on the surface of our planet, making measurements and obtaining images at visible and infrared frequencies. ASOS weather stations make very local meteorological measurements automatically; satellites make weather observations over a large swath of the globe automatically.

There are two types of weather satellites, distinguished by their different orbits and, as a consequence, different images. Equatorial satellites travel around Earth above the equator in circular *geostationary* orbits. The period (circle time) for such orbits is exactly one day, and a geostationary satellite rotates in the same direction as Earth. Consequently, the satellite appears stationary from an observer on Earth; it sits above the exact same spot all the time. One such satellite can therefore see less than half the globe, and so several are needed to cover the whole world. The period of a satellite depends on its orbital radius; those farther out from Earth orbit more slowly than those closer to the surface. A geostationary orbit has a radius of 42,157 kilometers (26,195 miles). That is, the satellite is 35,786 kilometers (22,236 miles) above the surface at the equator—about 5.6 Earth radii.

Polar satellites appear to move in a complicated manner above the surface. They orbit from pole to pole, making approximately 14.1 orbits per day. The radius for such orbits is about 7,200 kilometers (4,500 miles), or approximately 830 kilometers (515 miles) above the surface (the radius of Earth is less at the poles than at the equator, and so a circular polar orbit is a variable distance above the surface). In figure 5.6, the relative sizes of geostationary and polar orbits are apparent.

Because Earth rotates beneath it, a satellite in polar orbit moves relative to the surface—it views a swath of the surface that changes every day. The polar orbits of weather satellites are *sun-synchronous*, meaning that the orbital axis presents the same angle to the sun at all times of the year. This characteristic means that when the satellite passes over a given point on the surface, it does so at about the same time of day as it did the previous time it passed over this point. Why is this important? For an imaging satellite, this orbital characteristic is important because it means that a given point on the surface is imaged with the same

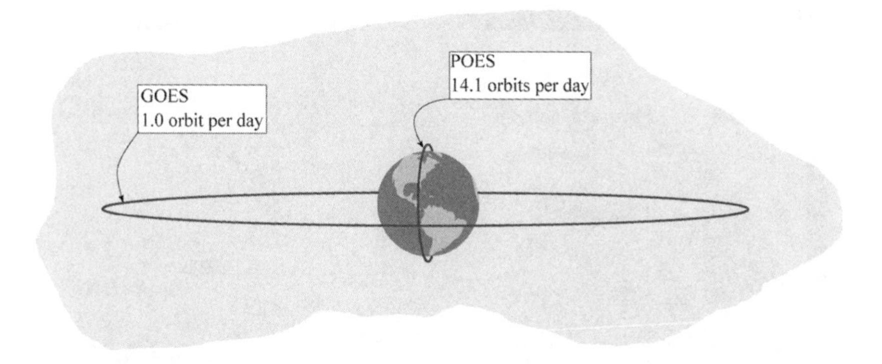

Figure 5.6 Weather-satellite orbits, shown approximately to scale. Polar orbits are at an altitude above the surface that is one-eighth of Earth's radius, whereas geostationary equatorial orbits are at an altitude that is about 5.6 Earth radii. Clearly, the surface coverage and data resolution of these two types of satellite are going to be very different.

angle of sunlight each time. That is, the surface is always seen in the same light, which makes for easier image comparisons.

From figure 5.6, it is apparent that the geostationary satellites can scan a large area of Earth but in less detail than a polar-orbiting satellite, simply because of the distances. The United States operates two equatorial satellites, known as Geostationary Operational Environmental Satellites (GOES). GOES-East sits at longitude 75°W (near the East Coast), whereas GOES-West is at longitude 135°W (off the West Coast). It also has two Polar Operational Environmental Satellites (POES).[7]

What are the pros and cons of the two types of weather satellite? Geostationary satellites provide wide-angle images (often shown on TV weather channels [figure 5.7]).They are good at providing information about large-scale, fast-changing weather phenomena such as storm development. (GOES transmits its data for the continental United States every 15 minutes.) They also act as data collectors, gathering information beamed up from weather buoys, balloons, and other spatially disparate data sources. But geostationary satellites provide images and data that are less detailed than those of polar-orbiting satellites. Furthermore, because of their equatorial orbits, they cannot image the poles. Equatorial-satellite data are updated at least every six hours; they tend to be used for short-term weather forecasts.

Figure 5.7 GOES-West hemispheric infrared image, showing water content of the atmosphere, January 15, 2015. GOES images can focus on regions of interest but not with the high resolution of POES images. (From National Oceanic and Atmospheric Administration)

Polar-satellite data are of higher resolution, though more limited in spatial extent (figure 5.8). The polar-satellite orbital characteristics means that the satellites cannot concentrate on a particular surface region of interest for very long; they snap and move, and no two images are the same. But they provide excellent images of the polar regions. Polar weather-satellite data are used for long-range weather forecasts. The "temporal resolution" (in plain-speak, the update rate) of polar-satellite data is poor compared with that of their geostationary cousins, but the spatial resolution is much better. The spatial resolution of POES data can be as little as 250 meters (820 feet).

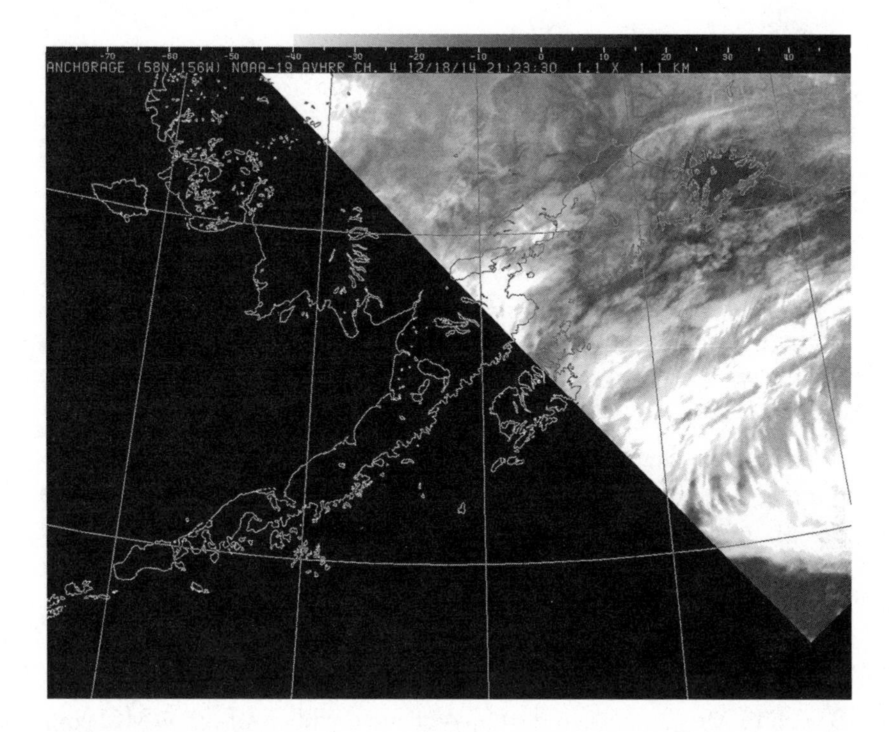

Figure 5.8 POES infrared image of southern Alaska, December 18, 2014. (From National Oceanic and Atmospheric Administration)

Satellite data cover the visible spectrum (daytime only, for measuring cloud cover, and ice and snow cover), and infrared (measuring cloud cover and altitude, land and sea surface temperature, nighttime fog distribution, atmospheric water vapor content, and movement).

Data Storage and Transfer

From this summary of weather data gathering, you may already have noticed a tendency for data sharing. Clearly, it is in the interest of different national meteorological groups to share data because the nations of the world share weather. Global weather prediction requires global weather data. How, in a world where international competition, suspicion, and enmity are not unknown, is this sharing achieved?

It is achieved through common interest, quite simply. We have seen that cooperation in weather-balloon release times permits a snapshot of the state of the atmosphere. Similarly, sharing weather data reduces the global cost of data gathering and permits meteorological data-processing centers in different countries to provide better predictions than they would otherwise. A further incentive to cooperation has arisen in recent years with the need to monitor climate changes; this factor increases the importance of both data quality and data quantity.[8]

The organized management of weather and climate data gathering extends to data processing. This organization has arisen slowly since the 1940s. First, local forms and procedures became standardized, facilitating the sharing of data, which became computerized from the 1960s. The World Meteorological Organization (WMO) has implemented standardized database management since 1985; this procedure was improved in the late 1990s for better access to the data by member organizations and to tighten up data security.

The WMO has member organizations from most countries. The U.S. representative is the National Oceanic and Atmospheric Administration; in the United Kingdom, it is the Met Office. The China Meteorological Administration, the Meteorological Service of Canada, the Australian Bureau of Meteorology, and so on—all are members of the WMO, and all contribute data and process the pooled global data that accumulates. From the China Meteorological Administration, we get a flavor of the workings of these agencies and of the WMO: "collection, processing, storage and retrieval services of . . . meteorological data . . . including global sounding data, surface observations, ocean observations, numerical forecast results, satellite data, soil moisture, sandstorm monitoring, wind data."[9] From the American Meteorological Society, we find a clear statement of the need for international cooperation: "The reliance of modern society on accurate weather forecasts for protection of life and property and as a basis of routine decisions has long been recognized and would not have been possible without the full, open, and timely exchange of data among the 191 WMO Member States."[10]

Around the world, there are many Regional Climate Centers, which collect, manage, and maintain large volumes of regional data. In addition, there are international data centers that maintain, develop, and homogenize high-quality global data sets.[11] Three such international centers are in the United States: the National Climatic Data Center and the World Data Center for Meteorology, both in Asheville, North Carolina, and the World Data Center for Paleoclimatology, in Boulder, Colorado.

The Hadley Centre for Climate Science and Services in Exeter, England, is another international data center; it holds complete sets of meteorological data that go back over 160 years (for example, global surface temperature).

There are, and always will be, challenges to the quantity and quality of the data that are gathered around the world. Of course, there are general measurement and sampling errors, and we investigate these and other statistical factors in chapter 6. In addition, there are inhomogeneities and systematic errors; for example, trees growing up around an observation station may change the light levels over a number of years. There is the problem alluded to earlier of scarce data over large areas of the planet surface—oceans, deserts, polar regions. Other problems that we have already encountered concern the need for data that extend in time as well as space: numerical weather prediction requires very accurate knowledge of current weather conditions around the world, but climate models require data (such as volumes and contents of volcanic ejecta) from other times if they are to reproduce observed effects accurately. We have also seen that some data are inherently more valuable than other data: where the dynamics leads to negative feedback, then initial data accuracy is not so important because errors will be wiped out, but where positive feedback reigns then initial errors accumulate—this effect is reproduced in spades in the chaotic regimes of weather dynamics, as we will soon see. An issue with data accuracy and completeness that is increasingly important concerns the size of weather cells used for numerical weather prediction and global-climate-modeling computer programs. Increasing prediction accuracy requires reducing the cell sizes; this obviously means more input data are needed, but it also means that new types of data are needed. This need for new data arises because new physics appears at smaller scales—for example, cloud formation.

There are statistical methods for minimizing the effects of imperfect input data. The implementation of these methods is a major player in the improvement over the past two decades of the accuracy of weather prediction. Forecasting is the subject of chapter 10, so I will say no more about it here. Now it is time to turn to the central role played by statistics (to many people the ugly duckling of mathematical sciences) in weather and climate.

***

Meteorological and climate data are gathered from the land and sea surfaces, from the troposphere and lower stratosphere. They are gathered by instruments that make direct measurements and by remote sensors, such as radars, and by satellites looking down at Earth from space. These data are gathered, transferred, stored, processed, and distributed by linked meteorological organizations around the world. The volume and accuracy of weather data are increasing, as is the type of data that are gathered. Statistical techniques are employed to handle inhomogeneities in the data and in general to minimize the effects of data imperfections.

6

Statistically Speaking

Statistics show that of those who contract the habit of eating, very few survive.
George Bernard Shaw

The subject of statistics is usually neglected in popular accounts of weather science, but it finds a place here because we are aiming for a deeper level of understanding. At this point, it is worth setting down my view, suggested in the introduction, that the essential ideas of statistics and weather physics can be conveyed without getting bogged down in mathematical details. True, for a polished theory of weather dynamics, you would need a textbook (coupled partial differential equations and all), but to gain insight all that you will need here are a few simple, well-chosen examples backed up with diagrams.

All those weather and climate data we saw in chapter 5 are rendered into predictions about future weather and climate by applying statistical methods of the type outlined here. But statistical concepts are more pervasive in weather dynamics and influence the very nature of what we can predict, as well as how accurately we can make the predictions.

Statistics Are Everywhere—Probably

The first notion that we need to fix firmly in our minds is the role that randomness plays in everyday life (and most certainly in the physics of our weather). We begin way out in left field, but you will quickly see where we are heading.

To date, five of the last seven U.S. presidents (from Gerald Ford to Barack Obama) have been left handed. The probability for any given male to be left handed is about 0.12, so if you were to pick a man randomly from the population, there is a 12% chance that he would turn out to be a southpaw. Here we have a common definition of probability—*frequency of occurrence*. For any given leftie (your author, for example), the fact that he turned out that way may or may not be due to chance. Who knows what goes on at the molecular level when an embryo forms in the womb? Whatever does happen includes handedness; we are born either left handed or right handed. Only 10% of females are left handed. Certainly, there is a genetic component to human handedness, but there is a larger environmental influence. Whatever the reason for individual handedness, we can use the mature science of statistics to analyze the handedness data, which brings us back to those presidents.

What are the chances that five of the last seven presidents would be lefties,[1] given the datum that 12% of males in general are so? A statistician would immediately refer you to the *binomial distribution*, which is the statistical tool for such a problem, and construct figure 6.1. Here we see that there is one chance in 2,500 that only two of the last seven presidents are right handed. This does not mean that we are the victims of a conspiracy (perhaps that should be "sinister conspiracy")—coincidences do happen—but it is suggestive of a preponderance of left-handed individuals in the field of politics. In fact, lefties do tend to prefer certain professions (engineering, in my case, is typical for a leftie).[2]

Different situations call for different statistical distributions. These distributions are derived mathematically from the circumstance of the situation being analyzed. The binomial distribution finds wide application in all kinds of unrelated fields:[3] gambling, genetics, sports, finance, and physics among them. Another that finds widespread application is the *Poisson distribution*, famously (among statisticians) used to analyze

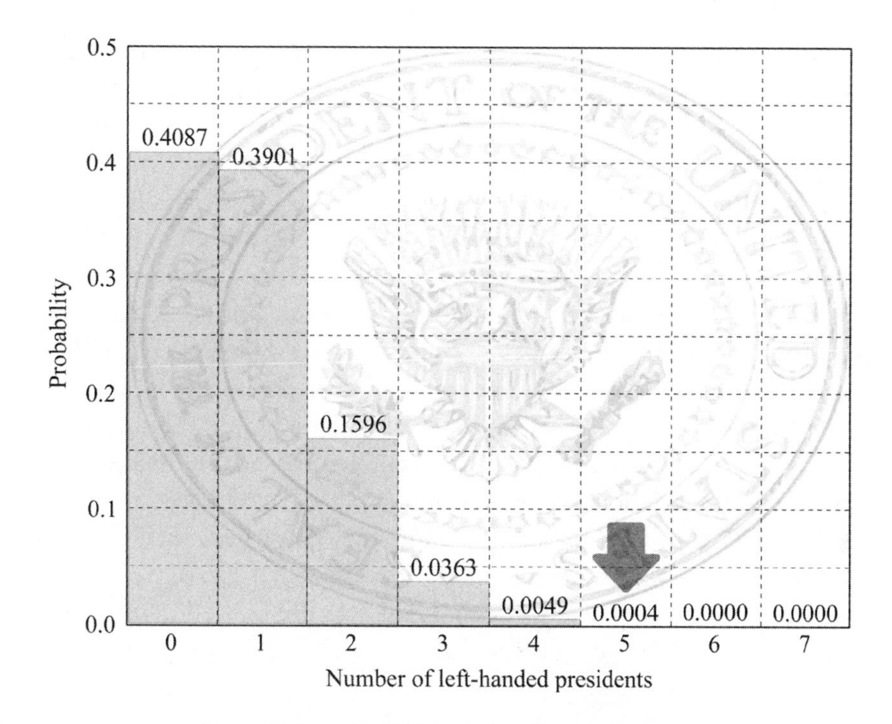

Figure 6.1 Presidential probabilities. The probability that none of the last seven U.S. presidents (Gerald Ford to Barack Obama) were left handed is about 41%, assuming a random distribution of handedness. The probabilities that one, two, or three of them were southpaws is about 39%, 16%, and 4%, respectively. In fact, five of the last seven presidents have been left handed. The chart shows that this is highly unlikely, based on national figures for male handedness, and so we conclude that (1) there is a *sinister* conspiracy or (2) the probability of left-handedness in American politicians is greater than that for Americans in general (if six times greater than five of seven lefties is the most likely number).

fatalities among nineteenth-century Prussian cavalrymen because of horse kicks. It seems that the Prussian army was alarmed by the numbers and wanted to know the cause. Analysis showed that the deaths from this cause per year between 1875 and 1894 were entirely consistent with random chance and, when plotted on a graph, closely followed the Poisson distribution. This distribution applies to many problems in telecommunications, physics, economics, and genetics.[4]

The Gaussian or *normal distribution* is ubiquitous. It is the iconic "bell-shaped curve" and applies so widely because many of the other distributions turn into the Gaussian under certain circumstances. Figure 6.2 shows but one application: the height distribution of Americans between the ages of 30 and 60 years, with Gaussian distributions superimposed. The average or *mean* height corresponds to the peak of the distribution (in this case, because the Gaussian distribution is left–right symmetric). The *standard deviation* refers to distribution width—the spread of heights. It is approximately the width of the distribution at half the maximum height, though the exact definition of standard deviation is more complicated and much more precise. We know that 68.2% of samples lie within one standard deviation of the mean, and that 95.4% of samples lie within two standard deviations ("2-sigma" or 2σ).[5] Thus the mean height of American men is 5 feet, 10 inches, and the standard deviation is 3 inches, and so we say *with 95% confidence* that an American man chosen at random will be between 5 feet, 4 inches and 6 feet, 4 inches tall.[6]

The recently discovered Higgs boson was announced at the particle physics facility of CERN (European Organization for Nuclear

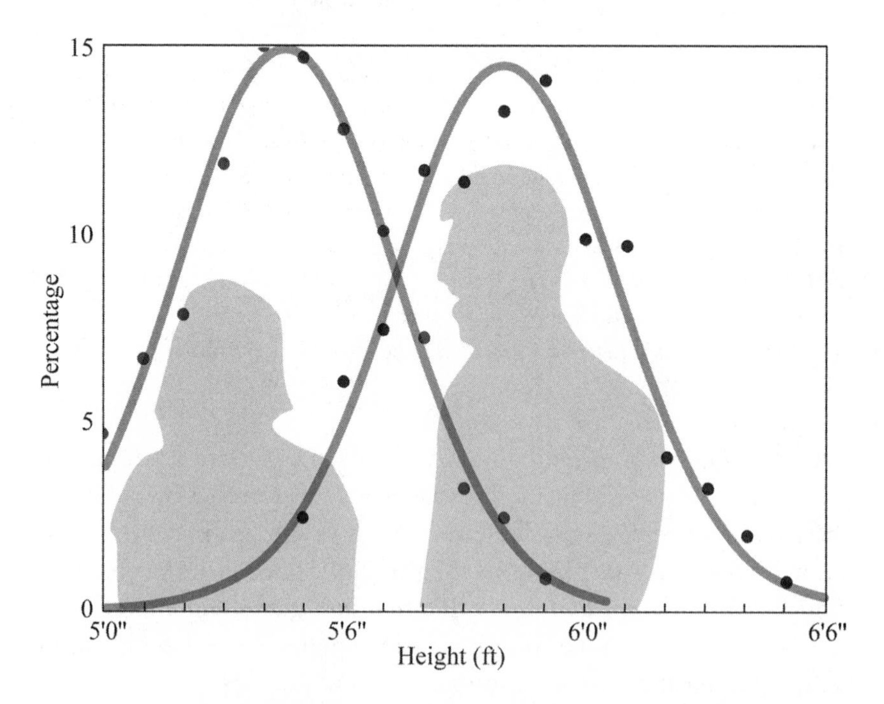

Figure 6.2 Standing tall. The Gaussian (*normal* or *bell*) curves for height distribution in American men (*right*) and women between the ages of 30 and 60 years.

Research) in Switzerland in March 2013. The criterion for discovery of a new particle is statistical: the confidence level must be 5-sigma, corresponding to one chance in 1,744,278 that its detection was due purely to chance. Statistics enters particle physics for the same reason it enters all of physics: because physics is about measurement, and measurement errors are statistical.[7]

Measurement Error

Suppose you measure the height of your child on her birthday. Let us assume that your tape measure is accurate, so there are no *systematic errors* in your measurement. The measure is marked off in tenths of an inch, and this means that you can accurately measure her height to the nearest tenth of an inch. A statistician would say that her height is, say, 4 feet, 6 inches "plus or minus a twentieth of an inch" or more scientifically 4.500 ± 0.004 feet (or even more scientifically, 1.3716 ± 0.0013 meters). The measurement error is taken to be half the minimum sensitivity of the measuring apparatus.

Weather and climate physics requires the measurement of huge amounts of data of many different types, as we saw in chapter 5. Temperature is measured with thermometers that may be graded to the nearest degree Fahrenheit[8] or centigrade; thus they inevitably incur a statistical measurement error. Pressure, air density, humidity . . . everything that is measured is measured to a certain accuracy and thus the measured value is one to which we assign a confidence level that is less than 100%. We may be 68% confident that the temperature at the bottom of your garden at 7:00 A.M. was 61°F (16°C), because 100 people with 100 thermometers were there, while you were indoors brushing your teeth, and they took 100 measurements and found a spread of values normally distributed about 61°F. The mean value was 61°, and all but five of the measurements were between 59° and 63°F (15° and 17°C) and so the standard deviation of the measurement is 1°. (Ninety-five percent of samples within 2° of the mean value corresponds to two standard deviations, and so $\sigma = 1°$.)

There are other sources of error than those due to measurement. Consider your young daughter. Her height changes with time of day (we are tallest when we first get out of bed, and compress throughout the day). When you measured her height on her birthday, was it always at the same time of day? If not, then her apparent growth rate, so lovingly measured, will only approximate her true growth rate. Also her mood may (almost certainly will) have been different on different days, influencing her posture. For any number of reasons, measurements introduce variability. We describe this variability, be it due to measurement error or something intrinsic to the phenomenon being measured, via statistics.

Starting Conditions and Chaos

A *deterministic* physical system is one that evolves over time in an entirely predictable way. For example, an ordinary pendulum starts at an angle of 20° (figure 6.3*a*), and the resulting movement can be predicted accurately by theory (figure 6.3*b*). The theory is not so simple when the pendulum start angle is large, say 70°, but the result can be calculated and, again, we can predict what the pendulum will do when released for times ahead that are much longer than the natural period of the pendulum. Thus if the pendulum is 1 meter (3.3 feet) long and its initial angle is 20°, then it will oscillate, as shown in figure 6.3*b*, with a period of 2 seconds. This prediction will hold valid for dozens of oscillations, even as the pendulum amplitude damps down due to air resistance (the change in oscillation amplitude is calculable if we have accurate knowledge of the aerodynamic drag of the pendulum). If the initial angle is 70°, then the period starts out at 2.2 seconds, and this gradually decreases to 2 seconds as the oscillation damps down—more complicated but still predictable and deterministic.

A problem arises when we take measurement error into account. Consider the pendulums that start off in the positions shown in figure 6.3*c* and *d*. The pendulum of figure 6.3*c* initially hangs vertically (start angle of 0°) and stays there as time progresses; it does not move—so far, so easy. The pendulums that started out at 20° and at 70° will both end up at 0°, due to air resistance, and so there is something about 0° that is special. We call it an *attractor*: the pendulum system is attracted toward this angle—obviously, this is due to gravity. Now look at figure 6.3*d*: here the start angle is 180°. This is also an attractor in the sense that a pendulum initially at 180° will stay at that angle; it won't oscillate, but

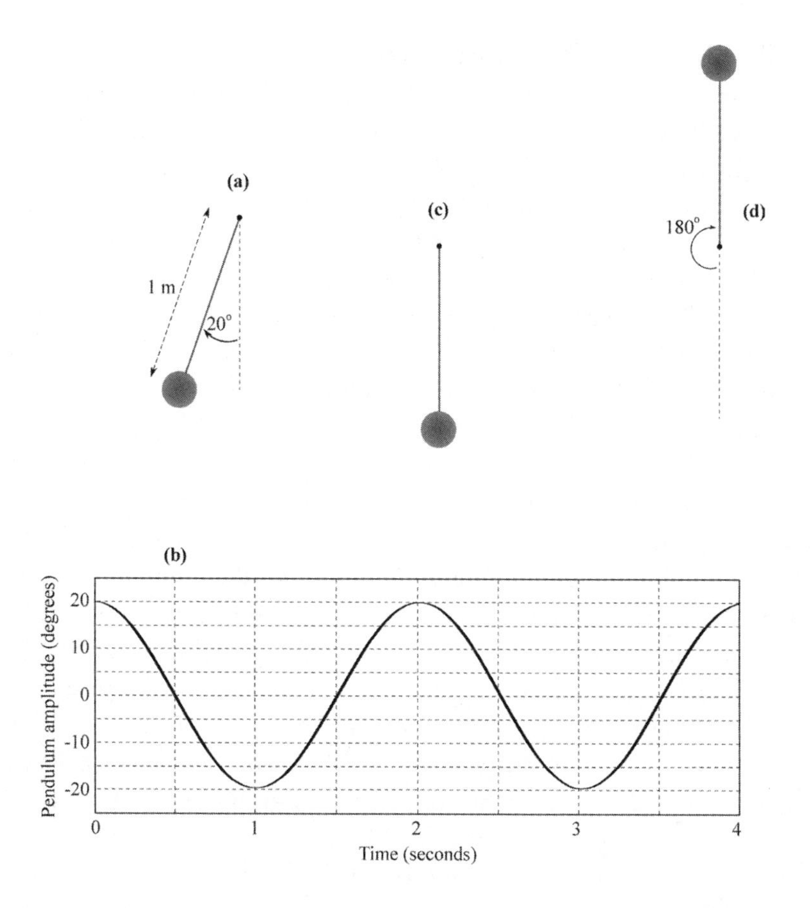

Figure 6.3 Pendulum problems. (*a*) A pendulum oscillates freely, starting at an angle of 20°. (*b*) The oscillation angle as a function of time; for this pendulum the period is two seconds. In principle, we can predict the motion of the pendulum far into the future. (*c*) If the pendulum starts out at an angle of 0°, then it stays there and, if nudged, will return there. This position is a stable attractor. (*d*) If the initial position is 180°, then the pendulum stays at that angle, but a small displacement causes it to move away from 180° with increasing speed (but which way?) This position is an unstable attractor.

will remained balanced like a pencil standing on its point. However, a small displacement from the attractor angle leads to very different results in the two cases. If you displace the pendulum of figure 6.3*c* by 0.4° and let go, it undergoes very small oscillations and soon ends up back at 0°. If you displace a pendulum by 0.4° from the vertical position of figure 6.3*d* (so that its start angle is 179.6° or 180.4°), the pendulum will undergo very large oscillations for a long period before eventually settling down to 0°. We say that the angle of 0° is a *stable attractor* for the pendulum system, whereas the angle of 180° is an *unstable attractor.* The pendulum in motion will head toward the stable attractor and away from the unstable attractor, driven by gravity. In fact, we can show theoretically that the pendulum initially at 179.6° or 180.4° will move away from the unstable attractor exponentially fast—picking up speed and accelerating away; this is characteristic of unstable attractors.

The problem in predicting pendulum motion arises because of measurement error. Suppose I set the pendulum initially at 180° as measured by a protractor marked off in 1° intervals. *Both* start angles mentioned earlier (179.6° and 180.4°) appear to be at the unstable attractor position according to my measurement, but the subsequent behavior is unpredictable because I do not know the start angle precisely enough. Suppose the pendulum really is at 180°; in this case, it will stay there for a long time. Suppose it is at 179.6°; in this case, it will start its trajectory by falling to the left. If it starts at 180.4°, then it will fall to the right. Here are three different behaviors for a very small difference in starting condition. If the small difference is less than measurement error, then we have a problem predicting what will happen. This simple example of a pendulum system exhibits one of the main difficulties that meteorologists have in predicting the development of our weather—a much, much more complicated dynamical system.

For the pendulum, the starting condition was a single parameter value—the start angle. Specifying this angle is enough to determine completely the current state of the pendulum. For our weather (and for climate, as we saw in chapter 4), there are millions upon millions of parameter values that have to be specified to provide us with precise knowledge of the current state of the weather, for prediction purposes. Among these parameters, we have to know the temperature, pressure, and humidity at every point of the atmosphere from ground level up to—at least—the troposphere. This is not enough; we also have to know about the state of the oceans. Obviously, it is impossible for meteorologists to know all that they need to know to make accurate predictions and so their predictions are necessarily approximate. For some aspects of weather prediction, the imprecise knowledge is unimportant. These aspects involve *negative feedback* and correspond to the pendulum dynamics that occur close to the stable attractor at 0°; we know that the pendulum will head toward 0° whether it started at 0.4° or at –0.5°. Other aspects of weather and climate physics involve *positive feedback* that corresponds to the pendulum near 180°; here imprecise data kill predictability. (We have encountered feedback earlier in this book and will do so again, more than once.)

The pendulum is a simple, deterministic physical system—it is hard to conceive of one that is simpler. Now let's look to the other end of the spectrum of physical systems. Unlike the simple pendulum example of figure 6.3, our atmosphere is a very complex system that exhibits *chaotic* behavior. "Chaos" is a word that has a special meaning to mathematicians, but its effect is quite close to the usual meaning. A chaotic dynamical system is one that evolves in time deterministically, like the pendulum, so that in principle we can predict far into the future what it will do if we know its current state exactly. However, a chaotic system is exquisitely sensitive to the initial conditions, much more so than our simple pendulum (at least we knew where it would end up, if not how it would get there). In figure 6.4, we see a simple mathematical example of chaos that illustrates a number of characteristics that a dynamical system must possess to be chaotic.

Consider a sequence of numbers [x_1, x_2, x_3 . . .] each determined by the previous, as follows:

$$x_{n+1} = x_n^2 - c \tag{6.1}$$

(Math-averse readers need not get alarmed: this is the only equation in the main part of the book.) Let us say that the constant c is given a value of 1.900000000 (as you will see, we have to be very precise) and that the starting value of the sequence is $x_1 = 0.500000000$. From equation 6.1, we easily calculate the next few elements of the sequence: $x_2 = -1.650000000$, $x_3 = 0.822500000$, $x_4 = -1.223493750$, and so on. Tiring of this mechanical calculation, I leave the next hundred elements of the sequence to an expert, my computer, and plot the result in figure 6.4*a*. The output (the sequence [x_n]) looks random. In fact, this is another characteristic of a chaotic system: the sequence is deterministic (for the same value of c and the same starting value for x_1 we will always obtain the same sequence) and yet it looks random.[9] A statistician analyzing the sequence generated by equation 6.1 would have to look hard to see that it is not random, but he or she likely could tell.

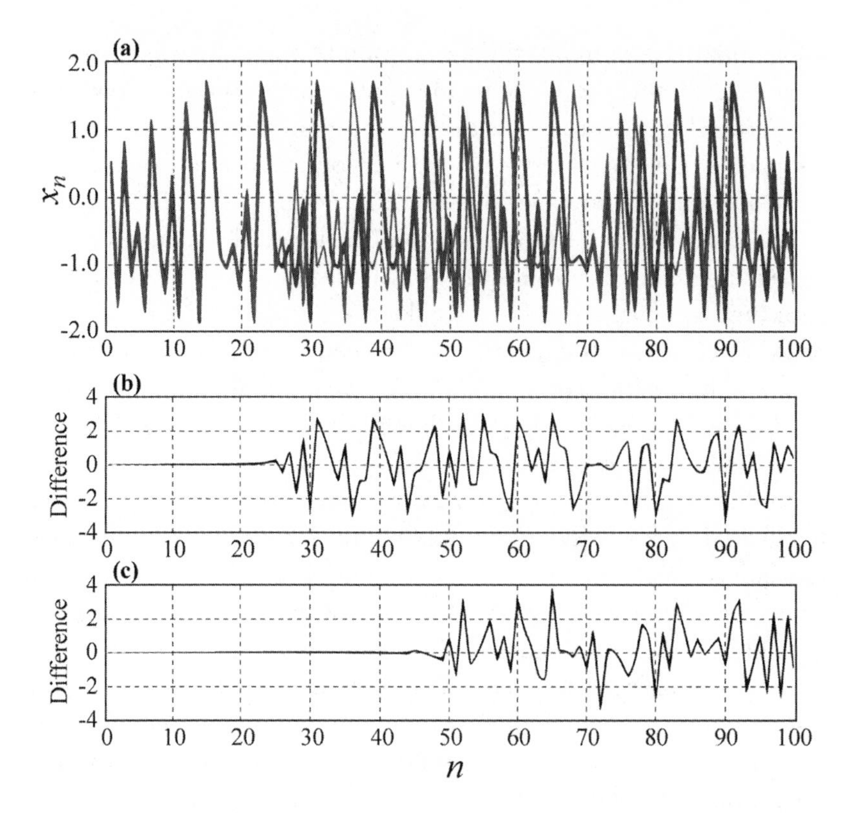

Figure 6.4 Chaotic sequence. (*a*) The sequence of numbers x_n, given by equation 6.1, versus *n*, for two starting values, $x_1 = 0.5000000$ and $x_1 = 0.5000001$. (*b*) The difference between these two sequences, which becomes significant for *n* greater than 24. (*c*) When the difference in starting value is reduced from a millionth to a billionth, the difference between the sequences becomes significant for *n* greater than 49. This observation tells us that, for chaotic dynamics, prediction requires very precise knowledge of initial conditions. Here, reducing uncertainty of initial conditions a thousand-fold leads to increased predictability of only a factor of 2.

Some deterministic sequences, however, look so random that they are statistically indistinguishable from genuinely random sequences in which the *n*th element is completely independent of the $n - 1$th element. (Such sequences are the product of *random number generators*—you have one on your computer.)

Now look what happens if we change the starting value of the sequence to $x_1 = 0.500001000$, without changing the value of *c*.

The result is shown in figure 6.4*a* and *b*. The difference between the two sequences starts out small but grows and grows until, for n greater than 30, the two sequences look completely different. Generating a third sequence, this time with $x_1 = 0.500000001$, yields the result shown in figure 6.4*c*: we have apparently unrelated sequences for n larger than 50. Suppose we didn't know the start value but had to measure it, and we measured it with an error of one part in a million. Figure 6.4*b* shows that we would be able to predict the sequence out to fewer than 30 samples; if our measurement error was only one part in a billion, then predictability would improve only somewhat, out to fewer than 50 samples.

Here we have another weather prediction dilemma in a nutshell—a storm in a teacup, as it were. Note that equation 6.1 is *nonlinear*, meaning that it involves the variable—here x—raised to a power not equal to 1, and that it involves feedback (the nth output becomes the n + 1th input). Both of these characteristics apply to weather system equations, in a big way. Note that not all nonlinear equations with feedback behave chaotically, but that all chaotic behavior originates from equations with these characteristics.

Weather behaves chaotically. Indeed, much of the initial insight into chaos theory came from meteorology, including a well-known popular label for the extreme sensitivity to initial conditions: the *butterfly effect*. This evocative name comes from a technical paper presented at a meeting of the American Association for the Advancement of Science on December 29, 1972, by American meteorologist Edward Lorenz: "Predictability: Does the Flap of a Butterfly's Wings in Brazil Set Off a Tornado in Texas?" Lorenz had written down a set of equations describing weather systems. These equations were nonlinear with feedback. They exhibited attractors. He was surprised to find that the computed solutions of the equations displayed extreme sensitivity to initial conditions (nobody knew much about chaos theory back then, though there were hints of its existence elsewhere in mathematics). Thus chaos and weather forecasting go back together a long way. (No doubt they will go forward a long way also, but in a manner nobody can predict . . .) Indeed, Lorenz came up with a pithy definition of chaos: "Chaos: When the present determines the future, but the approximate present does not approximately determine the future."[10]

Since its discovery, mathematicians and physicists have learned a great deal about chaos and have found that chaos theory applies to many aspects of the natural world. For example, it applies to the border region that lies between laminar (smooth) flow of a fluid and turbulent flow. Given that fluid flow is the essence of atmospheric physics, we can see why weather is chaotic and therefore why it is intrinsically unpredictable. For the very simple example of equation 6.1, we saw that, if we improve measurement accuracy by a factor of 1,000, then we improve the period of reliable prediction by less than a factor of two. This is the basic reason why we will never be able to predict our weather a month into the future.

Prediction, Amid Randomness and Chaos

To summarize quickly the results that we have found so far, concerning the limits to predictability, we note:

1. Prediction is limited by randomness (due to measurement error, for example).
2. Prediction is limited by imprecise knowledge of initial conditions, especially if the governing dynamics can exhibit chaotic behavior.

We might add a third point here:

3. Prediction is limited by our knowledge of the underlying dynamics.

That is, we don't know what is going on physically, or our theory about what goes on is wrong in some way. Naturally, such a circumstance would lead to incorrect predictions. However, and perhaps surprisingly, this point is less important than the other two. If our predictions are wrong because the underlying theory is wrong, then this fact quickly becomes apparent and we either fix the theory, if we know how, or we avoid making predictions in that area until we have a better theory. If, however, we have the right theory and yet we cannot make long-term predictions for the two reasons given, then what should we do? In this section, we give concrete form to this dilemma, again with simple examples, and show what can be done to mitigate the problems.

First, here's a no-brainer. Can you predict the next few terms in the sequence $x_1 = 1, x_2 = 2, x_3 = 3, x_4 = 4, x_5 = 5, x_6 = 6, \ldots$? (figure 6.5*a*).

There are no tricks here. I introduce this straightforward example of prediction to make a point. If there are no measurement errors (we have perfect knowledge of initial conditions) and if we have a good understanding of the underlying dynamics (we know that the sequence is a straight line), then it is trivial to predict ahead infinitely. We know what x_{999} is going to be.

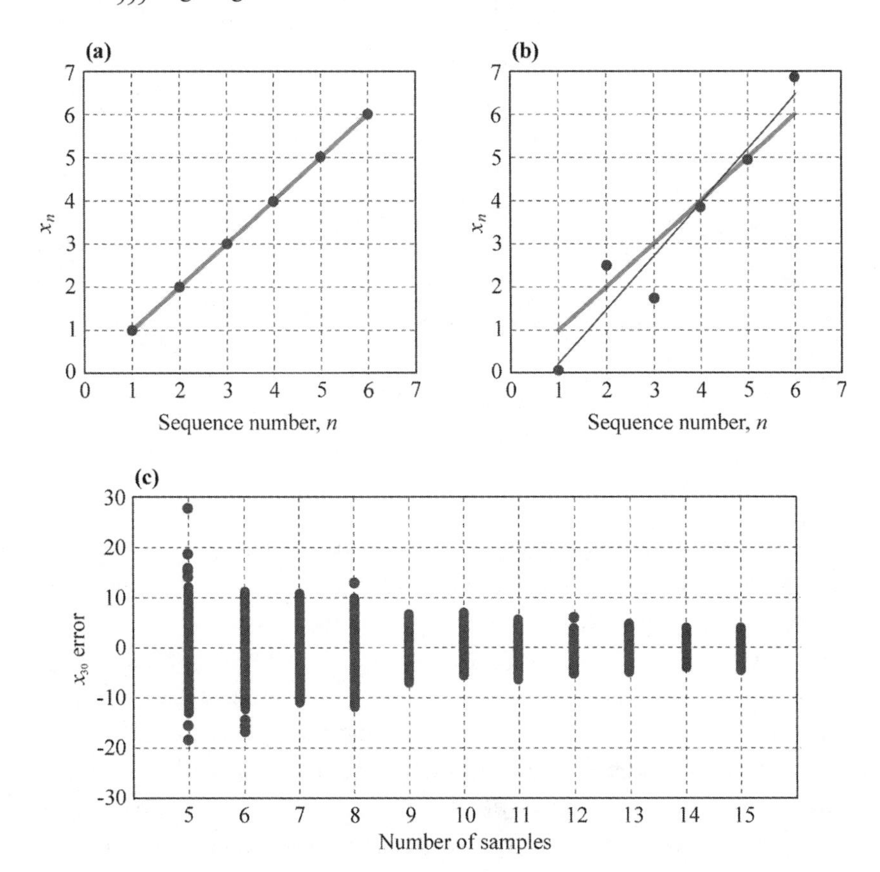

Figure 6.5 Measurement error and predictability. (*a*) A simple sequence of data (for example, temperature at the bottom of your garden at 9:00 A.M. on July 4 each year). If we knew these data exactly, prediction would be easy. (*b*) Measured values for the sequence (*dots*) plus the best-fit line for the data (*thin line*). Predictions based on this line will increasingly diverge from the real data sequence (*thick line*). (*c*) If more samples of data are measured, then prediction error is reduced. Here many different data runs, with random measurement errors, are utilized to show the difference between true and predicted value of x_{30}, as a function of the number of samples used to make the prediction.

But now let us introduce measurement error to this trivial example. Recall that measurement error is the first limitation to predictability on our list. Now even with perfect understanding of the simple dynamics, we can make predictions only with a given probability. The data shown in figure 6.5*b* incorporate measurement errors due to *white noise*, which obeys a Gaussian distribution with, in this case, a standard deviation of 1. There is a technique for optimally predicting the sequence for this elementary case, known to statisticians as *linear regression*. The technique is optimum in that it guarantees us that the prediction error is as small as it can be.

The predicted straight line (see figure 6.5*b*) differs from the true sequence line because of measurement errors, and consequently our estimates for x_7, x_8, x_9, and so on will be increasingly wrong. Stated more precisely: subsequent values of the sequence $[x_n]$ will be predicted with increasing error. We are able to calculate the standard deviation for this error and so can provide a warning on the can: "Consumers of this product should not trust it. We make it, and we don't trust it." This is hyperbole, of course. The statistical uncertainty of our prediction provides a measure for how much we can trust it; for example, we are able to say that x_7 will be within 1.45 of the true value 95% of the time. In fact, for the example of figure 6.5*b*, the predicted value for x_7 is too high, with an error of +0.76. The error in the predicted value of x_8 is +1.02, and extrapolation of our estimate further into the future (for $x_9, x_{10}, x_{11} \ldots$) yields ever-increasing errors. Thus our prediction is believable only for a sample or two ahead; we cannot believe the prediction for, say, x_{30}.

We can improve on our prediction by including more data points in the calculation, say by including x_1 to x_{15} and then predicting where x_{16}, x_{17}, and so on will lie. There will still be a prediction error, but it will be smaller because more data points leads to reduced error. We see in figure 6.5*c* how the errors in predicting x_{30} decrease as we increase the number of samples utilized in making the prediction from x_1 to x_5 to x_1 to x_{15}. How does the point of this little exercise relate to weather forecasting? More samples mean reduced errors, and so we expect that predictions made over large areas of the globe will contain smaller errors than predictions made over small areas. Thus we may accurately forecast next Tuesday's mean temperature for Washington State, but our prediction for the mean temperature of Seattle will likely have a larger error.[11]

In fact, the empirical predictions of the kind just discussed, and errors associated with them, are only a small part of the statistical issues that arise with weather forecasting. The effects of chaos and of imperfect initial data are much more pervasive, because we understand the dynamics of weather evolution but do not have so precise a knowledge of the starting conditions. Faced with these consequences of measurement error and of the intrinsic chaos of weather physics, what can meteorologists do? Their job is to forecast the weather, and so they must present us with such predictions, errors and all, along with weasel words as to prediction accuracy. We are not told that tomorrow will bring us 1.2 inches (30 millimeters) of rain, but are told that the probability of precipitation (POP) for this amount is, let us say, 40%. A useful technique for managing the sensitivity to initial conditions is *ensemble forecasting*. Once the role played by chaos became widely appreciated, in the 1980s, ensemble forecasting became the norm; today, all the main meteorological centers around the world and research facilities in universities apply this technique. The idea is to adopt not just one, but many sets of initial conditions. So if your measurements tell you that the temperature at the bottom of your garden at 7:00 A.M. today was 61°F, you put that value into your forecasting model (the set of equations that you use to predict weather) along with a bunch of other values that are close to the measured value (say 60.0°F, 60.1°F, 60.2°F, . . . 62.0°F). By adopting a spread of initial values for all the input parameters of your model, you hope to include the true values and so make predictions that represent what is actually going to happen. If you are confident in a particular parameter measurement (small measurement error), then you need include only a small range of initial values for that parameter in your ensemble; if you are less sure, then you would include a wider spread.

We can see the effect of ensemble forecasting by returning to the simple chaotic system of equation 6.1 and figure 6.4. Suppose—just for the purpose of illustration—that the Canadian Meteorological Center decides that the mean temperature in Flin Flon, Manitoba, fluctuates daily according to equation 6.1. It measures the mean temperature today and predicts what it will be for the next few weeks. Knowing about ensemble forecasting, the Canadian Meteorological Center spreads it bets by including a range of mean temperatures that are close to the measured value. Let us say that it measures a mean temperature that is 0.500000°C above yesterday's mean value. The ensemble includes temperature changes in the range 0.499995° to 0.500005°C, and the resulting prediction is shown in figure 6.6 (compare figure 6.4).

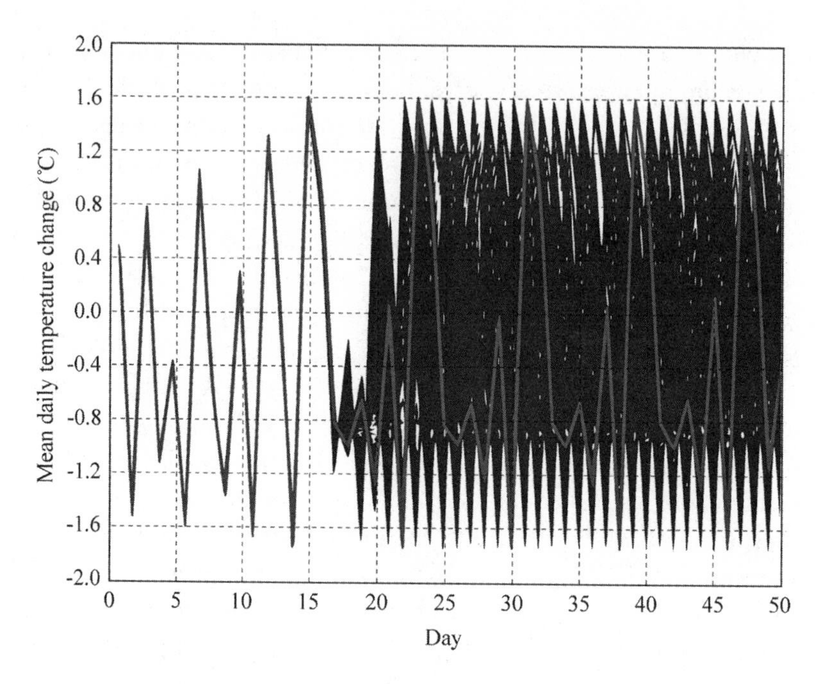

Figure 6.6 Ensemble temperature-change predictions (and actual temperature changes) for a simple weather system governed by equation 6.1. For this system, sensible predictions can be made 17 days into the future.

The ensemble forecast shows that mean temperature change can be forecast quite well for the next couple of weeks, in this hypothetical case, but after that it quickly becomes meaningless; the different initial conditions lead to very different predictions. Thus ensemble forecasting provides an indication of just how far into the future your weather model works. In a real system, the model depends on many parameters, and the predictions are better for some areas and some conditions than for others, but the idea is the same.

Ensemble forecasting is a type of *Monte Carlo* analysis. Monte Carlo methods are named after the famous Mediterranean gambling casino because they use chance—random numbers—in selecting (in our case) the spread of initial values. These values are chosen based on the estimated likelihood that they are correct, thus weighting the selection of parameters that are chosen as input to the weather-forecasting model. It may seem strange to introduce blind chance into the proceedings, and indeed that would be strange, but here the chance is not blind.

The statistics of parameter selection represent our (imperfect) knowledge of the true state of the weather at a given instance of time, and so by applying statistical techniques to the process of selecting parameter values, we are inputting information, not blindly spreading a net and hoping to catch a break.

To illustrate the efficacy of Monte Carlo methods, we here apply the Monte Carlo notion to a fully deterministic problem: estimating the area of an ellipse. In fact, we know how to calculate this area exactly (it is πab, where a is half the longest axis of the ellipse and b is half the shortest axis). We estimate the area as described in the caption to figure 6.7. The point is that the Monte Carlo method can be utilized to

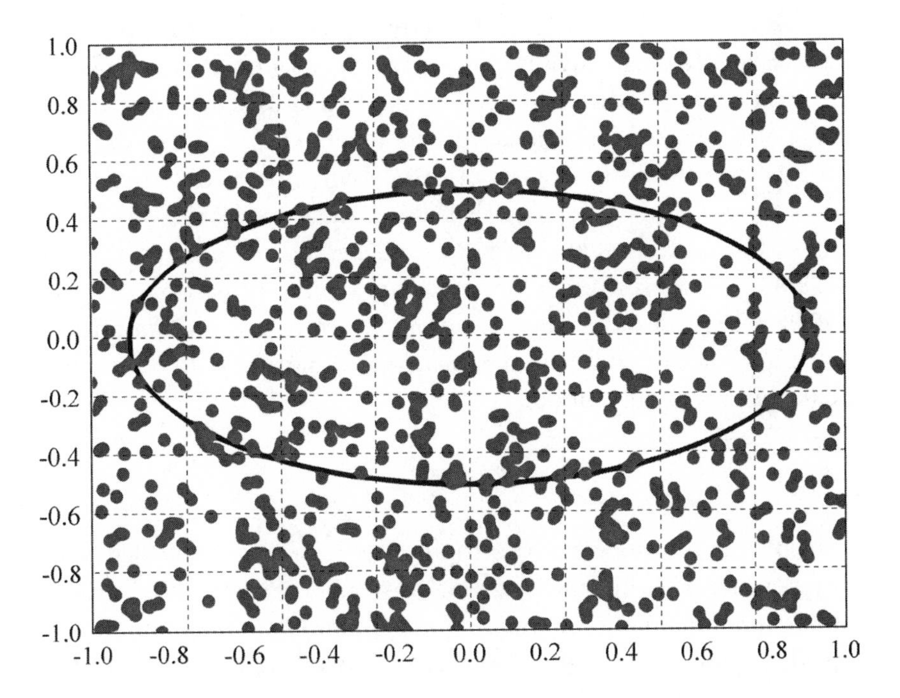

Figure 6.7 Monte Carlo method. We can use random numbers to help us with calculations that have nothing random about them. Here the ellipse covers an area of 1.413716694 . . . We fill the square with 1,000 randomly placed points, and count how many are inside the ellipse. Do this 10 times to obtain an estimated area of 1.386 ± 0.044. Do it 10 times for 100,000 points, and our estimate improves to 1.4123 ± 0.0057. The advantage of the Monte Carlo approach is that it works just as well for much more irregularly shaped areas, or volumes, which other estimation methods find difficult.

find any area or volume, not just areas and volumes where a formula is known to apply. For example, let us consider your daughter again. You decide, for reasons best known to yourself, that you need to know her volume. So in addition to estimating her height on each of her birthdays, you decide to measure her volume. One way would be to follow Archimedes and dunk the poor girl in your bathtub and then measure the volume of displaced water, but it's not an option in this situation. So you scan her into your computer and estimate her volume with the Monte Carlo method of figure 6.7.

Note that the result of a Monte Carlo calculation has an error associated with it—naturally enough, given its statistical nature. For a deterministic problem such as estimating the area of an ellipse, this is an avoidable overhead, but for weather dynamics it is not: such processes will have error bars anyway. Yes, weather is chaotic and so is fundamentally deterministic, but because we do not have a perfect knowledge of the parameters, it behaves very much like a random system. A significant part of the progress made in weather prediction over recent decades has been in the way that errors and uncertainties have been managed, so that we know what confidence we can place in the predictions of our weather models. Another significant part of our progress has been in reducing the size of the errors and uncertainties by obtaining data of greater accuracy, and much more of it, to be number-crunched by bigger computers.

***

The physics that governs weather processes is chaotic, hence deterministic in principle, but the evolution of weather is not deterministic in practice because we have imperfect knowledge of initial conditions. To predict tomorrow's weather, we need complete and perfect information about today's weather. Inevitably, the data we have about current weather conditions are such that prediction errors are unavoidable, especially given the chaotic nature of key weather phenomena. Ensemble forecasting helps provide a statistical context for predictions.

7

A Condensed Account of Clouds, Rain, and Snow

It always rains on tents. Rainstorms will travel thousands of miles, against prevailing winds for the opportunity to rain on a tent.

Dave Barry

Without water, our weather and climate would be unrecognizable. The importance of water—be it solid, liquid, or gas—for weather systems is emphasized here. We look into the causes and types of cloud and of precipitation.

Clouds Are Crucial

You know already the basics: clouds are aggregates of water droplets (or ice particles) in the atmosphere that appear white or gray, or perhaps red, orange, or yellow, depending on sunlight. You may not know that individual clouds can weigh thousands of tons, and that Earth's cloud cover influences our climate profoundly. You might expect clouds to take on an infinity of forms, and yet the various shapes and colors permit classification and the different classes of cloud have different consequences for both weather and climate.

The crucial role played by clouds in the energy balance of our planet arises because clouds can either heat or cool the world, depending on their altitude, size, and composition. A cloud cools the world beneath it by reflecting incoming solar shortwave radiation back into space, so that the surface beneath the cloud absorbs less solar power than it would if the cloud was not there. But a cloud can also warm the world by trapping longwave radiation that is emitted by the planet surface. Generally, low and thick clouds (such as stratocumulus)[1] contribute to planetary cooling, whereas thin, high clouds (cirrus) warm the planet. The overall effect of all the clouds that are present at any given instant of time (which cover, typically, 60% of the globe) is to somewhat cool Earth; cooling dominates because clouds reflect 20 to 30% more sunlight on average than does the surface. If you think back to the discussion of blackbody radiation and the greenhouse effect in chapter 1, and consult figure 7.1, you will see why high and low clouds influence the global energy budget in the way that they do.

Clouds form when the water vapor content of the atmosphere exceeds the maximum amount that is can hold—that is, when the air becomes saturated, a phenomenon that is strongly temperature dependent (details are in chapter 8). Colder air can hold less water vapor than warmer air, so air that cools as it rises reaches a certain height at which vapor condenses out into visible droplets of water. Thus from an airplane window we sometimes see clouds at a very specific altitude—they look like cotton balls on a sheet of glass—and this phenomenon reflects the fact that air temperature falls with increasing altitude. The reverse process also occurs: clouds dissipate when air temperature rises, causing water molecules to evaporate off droplets and return to gaseous water vapor. A cloud that is not growing or dissipating is in equilibrium: the evaporation rate of water within it equals the condensation rate of water vapor as individual water droplets grow and shrink.

The atmosphere become saturated with water vapor for one or both of the following two reasons: the local water vapor content increases (due to evaporation from the ocean surface, for example), or the local air temperature falls (due to the air being forced upward as it blows against a mountain, or because the sun sets, for example). But to form clouds, saturated water vapor in practice needs a *condensation nucleus* on which to develop a droplet. This nucleus may be a microscopic speck of dust, a salt crystal, pollen, a liquid aerosol, or a solid pollutant such as soot from a smokestack.[2] Its substance is *hygroscopic*—it

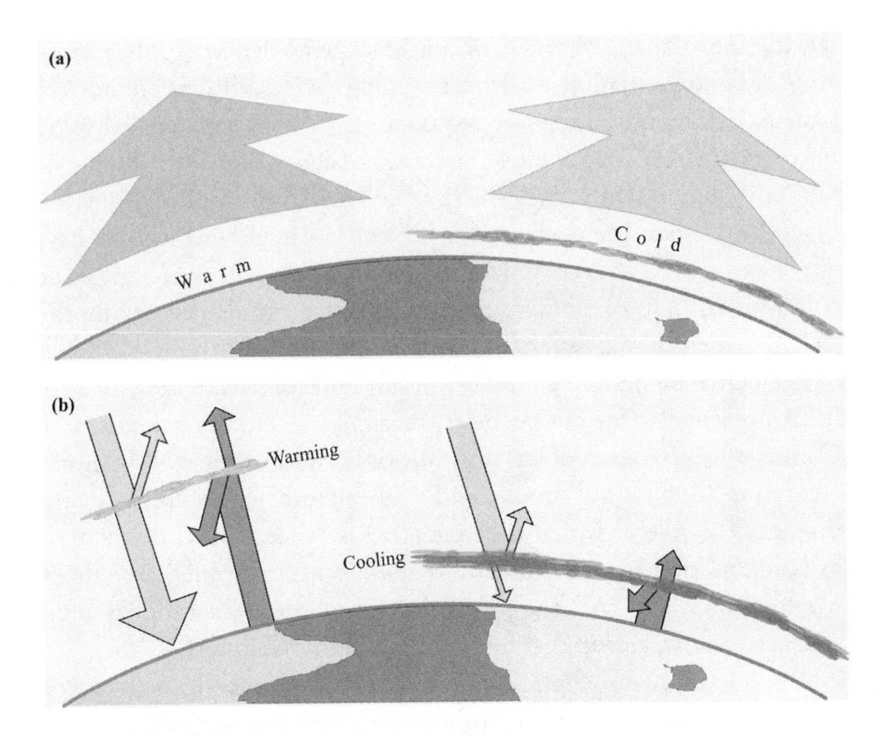

Figure 7.1 Clouds and Earth's energy budget. (*a*) The tops of clouds are cooler than the surface of Earth—by 19°C, on average—and so the blackbody (longwave) radiation temperature is lower. Thus clouds can help retain solar power. (*b*) In more detail, high clouds transmit most incoming shortwave radiation but reflect or absorb most outgoing longwave radiation and so warm Earth. Low clouds, though, are denser and so transmit less incoming shortwave radiation while transmitting longwave radiation more or less equally in both directions. Thus the overall effect of low clouds is to cool Earth. On balance, the cooling effect dominates.

attracts water. If no such nucleus is present, then the air can become supersaturated. The need for nuclei doubles at high altitudes, where the air temperature is below the freezing point. If no condensation nuclei are present, then water vapor cannot condense to form water droplets; even if there are such nuclei and water droplets form, they may not be able to freeze without *ice nuclei* being present. Thus supercooled liquid water at 10° or 20°C (18° or 36°F) below freezing temperature can persist until a suitable substrate appears (such as an airplane, which ices up quickly when flying through clouds of supercooled water droplets).[3] Note that condensation nuclei for seeding water droplets are not the same as the ice nuclei that are required for turning supercooled water into ice particles. The ice nuclei need to be shaped like natural ice crystals, and consequently the density of ice nuclei in the atmosphere is much lower than the density of condensation nuclei, which can have any shape. This is the reason why clouds of supercooled water are quite common at high altitudes—supercooled water clouds become mostly ice only when they are very cold (–40° or below).[4]

Condensation nuclei are small—typically 1 micrometer (0.000039 inch) in radius (smaller for salt particles, more like 0.1 to 0.5 micrometer). Consequently, water droplets start out tiny, and grow to a 10-micrometer radius for a typical droplet and a 50-micrometer radius for a large droplet. The rate at which droplets grow is faster than could happen by condensation alone: turbulent air conditions in clouds cause droplets to collide and coalesce—typically, a million small droplets must coalesce to form a drop that is large enough to precipitate.

One of the many complications of climate modeling is that relevant phenomena occur on many different length scales, all of which need to be modeled physically for predictions to be accurate. Thus tiny condensation nuclei influence the global formation, distribution, and characteristics of clouds, which in turn influence the results of climate models (and of numerical weather predictions). Why is this so? We observe and expect that tropical zones are cloudier than subtropical and polar zones (by 10 to 20%) and that tropical cloud tops are higher (by 1 to 2 kilometers [0.6 to 1.2 miles]), because of the warmer air in tropical regions. Thus atmospheric temperature distribution influences cloud characteristics globally—no surprise there. But how can cloud nuclei influence cloud characteristics on a global scale? Condensation nuclei are more common over land than over the oceans, and this fact contributes to the difference between maritime and land clouds. It means that "dirty" clouds over land must be deeper than "clean" clouds over the ocean before they initiate precipitation.[5]

Different thermal properties of land and oceans also contribute to these differences between clouds over these regions:

1. Oceans are significantly cloudier than land (67% versus 50%, including 50% dense clouds versus 15% dense).

2. There are more ocean clouds in the morning, and more land clouds in the afternoon.
3. Ocean clouds reflect 10% less sunlight than the same area of land clouds.

Now that we have seen how clouds affect the global absorption of energy from the sun, and how the global distribution and character of clouds vary, it is time to turn to the different types of clouds that meteorologists recognize.[6]

Clouding the Issue

It may seem strange that the classification of clouds was not one of those prescientific developments that came down to us from antiquity, like our classification of materials (rock, water, air, fire) or of bodily fluids or humors (black bile, phlegm, blood, yellow bile). It seems positively perverse that clouds were not placed into a rational categorization until after Linnaeus, who in the eighteenth century did just that for the much more complex, mutable, and wide-ranging collection of animals and plants that populate our planet—yet it is the case. Luke Howard, a pharmacist and an amateur meteorologist, gazed up at English clouds in the first decade of the nineteenth century and wrote about them. The same observations could have been made centuries earlier because they did not depend on any modern scientific knowledge or equipment. Despite this lack of technical rigor, the classification of clouds that Howard first set down in 1803 remains largely intact, unlike the early classifications of materials and humors.[7]

Howard designated three basic forms—the building blocks for all clouds—plus one special form for rain clouds, which today we combine to form 10 basic cloud types. The four basic forms (referred to as "the core four" by the National Weather Service) are

1. *Cirro-form* are high, wispy ice clouds, resembling curls of white hair. They appear in advance of a low-pressure area, such as a mid-latitude thunderstorm or tropical hurricane. They are made of ice crystals.

2. *Cumulo-form* are dense, white, fluffy cotton balls that are well defined and often with a flat base. They develop vertically and can be the tallest of clouds, due to strong convective updrafts. They are made

mostly of liquid water droplets that form as a result of convection and are associated with summertime afternoon convection.

3. *Strato-form* are broad, flat blankets often merging or with diffuse edges. They develop from nonconvective rising air—warm fronts, convergence zones, up-slope flows. They are associated with liquid water droplets, of diameter 10 to 30 micrometers (0.00039 to 0.0019 inch).

4. *Nimbo-form* are gray rain clouds showing elements of the other three forms.

Ten basic cloud types, formed by combining these forms, are recognized today (figure 7.2):

1. *Cirrus* appears as wispy white filaments that are lit up by the sun earlier and are the last to disappear at night, because they are so high (figure 7.3). Before sunrise and after sunset, it often appears brightly colored—yellow or red. These clouds are harbingers of bad weather because they develop in front of mid-latitude cyclones.

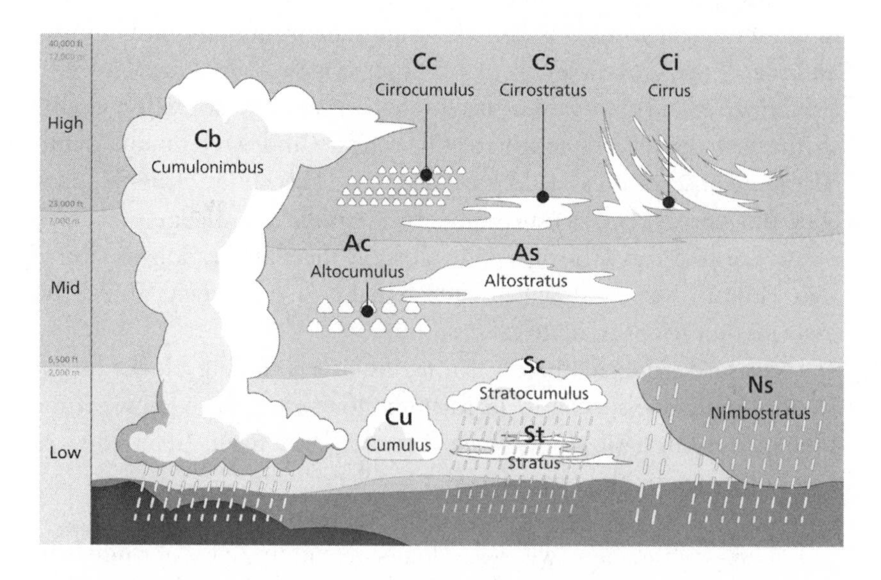

Figure 7.2 The 10 basic cloud types. They are categorized by altitude, so note, for example, the progression from cumulus through altocumulus to cirrocumulus. (Valentin de Bruyn, "Cloud Classification," Wikimedia Commons, http://d3j245lxrdmehy.cloudfront.net/0/?url=Z3ZzLm5lX3NlcHl0X2Rlb2xDQTMl ZWxpRi9pa2l3L2dyby5haWRlcGlraXcubmUvL0EzJXNwdHRo)

2. *Cirrocumulus* is an uncommon cloud forming a layer of thin, uniform sheets. It is seen in different combinations with other cloud types, indicative of continued fair weather or of rain to come in the next few hours.

3. *Cirrostratus* is a large transparent veil. The sun and moon generate halo effects when seen through these clouds. When increasing across the sky, it is indicative of fair weather changing to rainy or snowy conditions—an approaching weak front.

4. *Altocumulus* is the most common mid-level cloud, appearing as an extended layer or layers of white-gray rounded masses or rolls, sometimes described as a "mackerel sky" (figure 7.4). It often appears with other cloud types, between warm and cold fronts.

5. *Altostratus* appears as widespread sheets of uniform gray-blue cloud, thin enough to show the sun but too thick to show shadows. These clouds do not produce a halo. They tend to form ahead of warm or occluded fronts.

6. *Nimbostratus* is a dark gray, widespread and uniform layer thick enough to blot out the sun, and generating precipitation.

7. *Cumulus* appears as dense white-gray cauliflowers with well-defined edges that are often white (sunlit) and rounded on top while gray and flat at the bottom (figure 7.5). Over land, it often develops during the day as the land heats and dissipates toward evening. It is usually indicative of fair weather and of thermal columns.

8. *Cumulonimbus* is the thunderstorm cloud, gray and dense, forming a huge mountain or tower with the upper part usually flattened into an anvil. Rain and hail fall from the base of these clouds. This is the cloud of cold front squall lines, as we shall see.

9. *Stratocumulus* appears as honeycombed sheets or layers of whitish gray, spreading over a wide area of sky. It produces corona coloration effects around the moon and is often associated with dull weather.

10. *Stratus* is a gray, uniform, and often thick layer spread across the sky—high-flying fog (figure 7.6). It indicates a stable atmosphere and does not usually produce a halo.

Meteorologists subdivide these basic clouds; there are photographic databases with many examples of each type and subtype of cloud, to aid identification.[8] What is the purpose of all this considerable effort? It is not just an expression of the organizing, categorizing

Figure 7.3 Wispy cirrus clouds, developing into cirrocumulus (*top right*). (Photo by Simon Eugster)

Figure 7.4 Altocumulus clouds. Note the mackerel pattern, which is sometimes called "altocumulus undulatus" by meteorologists. (Photo by the author)

Figure 7.5 Cumulus clouds, typical of trade-wind latitudes, September 21, 2012. (From NASA, Earth Observatory)

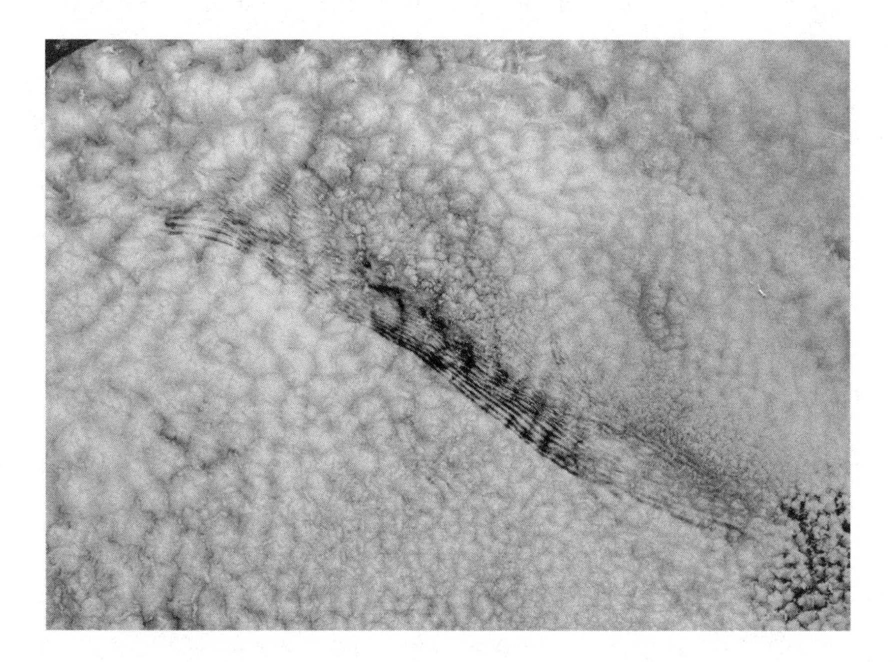

Figure 7.6 The top of stratus clouds located off the coast of West Africa, in the South Atlantic, August 28, 2003. From below, stratus clouds are much less interesting looking: a uniform sheet of gray. The area covered by this image is 550 by 425 kilometers (340 by 265 miles). (From NASA's *Aqua* satellite)

urge of enthusiastic backroom scientists (okay, nerds); knowing the types of clouds that occupy the sky at a given instant and how these types are changing tells meteorologists about the state of the sky and how it (and the weather) is going to develop. For example, the density of clouds and the size of their water droplets depend on cloud type (figure 7.7). Think of clouds as horses in a race and meteorologists as punters placing bets on the outcome. Identifying the leading horses and knowing something about their individual characteristics enhances the likelihood of picking a winner—that is, of estimating future developments.

The cloud menagerie includes stranger and more unusual beasts such as the *lenticular* (figure 7.8) and the *mammatus* clouds (figure 7.9). There are rare and beautiful *nacreous* clouds way up in the lower stratosphere (they are very cold, below –78°C [–108°F]), often seen over polar regions lit from below when the sun has just set. Another rarity is the *Kelvin–Helmholtz* cloud, which forms an extended, billowing wave pattern and that arises from the turbulent flow that is a consequence of strong vertical shear between two air streams. The formation, composition, and meteorological significance of most of these

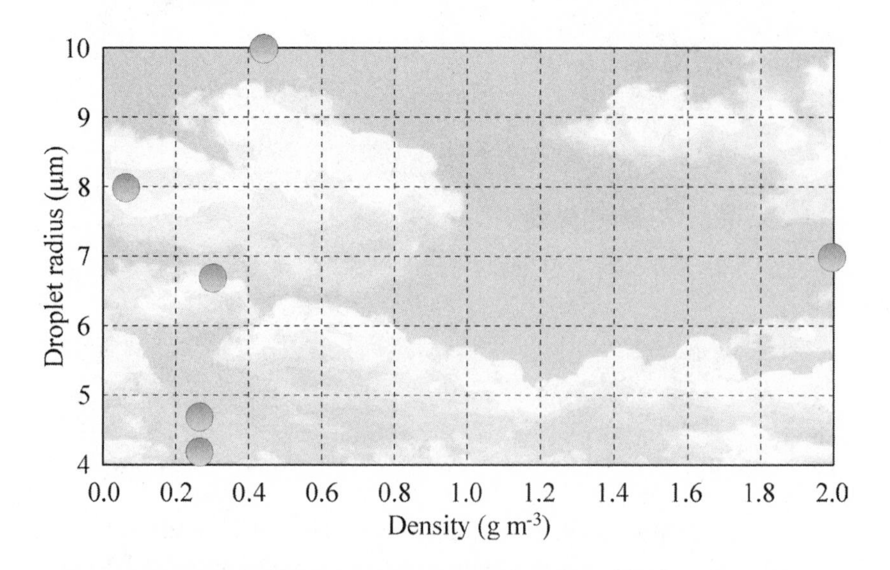

Figure 7.7 Average water-droplet radius versus cloud density for some common clouds over land and oceans. The variation in droplet sizes is large.

Figure 7.8 The oddly shaped lenticular clouds, forming over a mountain in Iceland, June 26, 2005. Sometimes these clouds are mistaken for UFOs. They arise from moist air being lifted over mountains. (Photo by Josvandamme)

Figure 7.9 Another oddity: mammatus clouds over Regina, Saskatchewan, Canada, June 26, 2012. They hang beneath cumulonimbus clouds and presage a thunderstorm. These are well lit from the side, showing their form. (Photo by Craig Lindsay)

clouds—common and rare—are well enough understood by science. Thus, for example, we know that stratus and stratocumulus clouds occur at low levels in the atmosphere and that they and all higher-level clouds are associated with low vertical winds speeds. This contrasts with cumulonimbus clouds, which spread through all heights and are associated with high vertical wind speeds (up to 40 meters [130 feet] per second), as we will soon see.

Knowing the cloud types that are observed in the sky, a meteorologist infers the atmospheric conditions that gave rise to them. Thus surface heating leads to thermals (convective heating) and so to cumulus clouds. Topography deflects winds upward producing layered clouds. Fronts, convergence zones, turbulence—all result in moist air being elevated in particular ways that lead to particular clouds.[9]

Fog

Fog is a cloud that reaches to the ground—nothing more or less. The generation of fog differs from that of clouds, however, because of the proximity of the surface. Whatever its origin and wherever it is found, if visibility is less than 1 kilometer (0.6 mile), then ground clouds are called *fog*; if visibility exceeds 1 kilometer but is less than 5 kilometers (3 miles), then they are *mist* (box 7.1). Fog is thickest in industrial areas where condensation nuclei near the ground are plentiful.

Fog is generated from a local water source such as a lake that is immediately beneath it. *Advection fog* appears as moist air passes over a cooler surface. The air cools and water vapor condenses, as for clouds. Such fog is most common at sea when moist air over warm water is then blown over cooler water (figure 7.10). In fact, the foggiest place on Earth is the Grand Banks—the (now depleted) fishing grounds off the coast of Newfoundland where the cold Labrador Current meets the Gulf Stream. Over land, advection fog appears when a warm front passes over snow.

Radiation fog often forms on calm nights in winter as the land radiates heat and so cools, thus cooling the moisture-laden air above it. If this cooler air then descends to lower lying land such as a valley, it can become very dense (and deep, up to 300 meters [1,000 feet]). Radiation fog dissipates quickly when the sun rises and heats the air

Box 7.1
Visibility

Visibility through air that holds water droplets depends on the size of the water droplets. Consider box figure 7.1*a*. Two droplets occupy a certain volume of air. In box figure 7.1*b*, the same mass of water occupies the same space but now the droplets are half the diameter of those in box figure 7.1*a*. Thus there are eight times as many droplets, each with one-quarter of the projected area of the larger droplets, so that the total projected area of water droplets is twice what it was before. From this simple geometrical argument, you can see that the visibility will be lower through air containing the smaller droplets than through air containing the same mass of water as larger droplets. In detail, we work out in the appendix the *mean free path* (the distance we can expect to see) through air containing water droplets; this distance turns out to be proportional to droplet diameter. Thus, for example, we can see through rain a distance of 1.3 kilometers (0.8 mile) if the raindrops have a diameter of 4 millimeters (0.16 inch) and the density of atmospheric water is 1 gram per cubic meter, whereas we can see only 3 meters (10 feet) through fog that has water droplets of a diameter of 10 micrometers (0.00039 inch) with the same atmospheric water density.

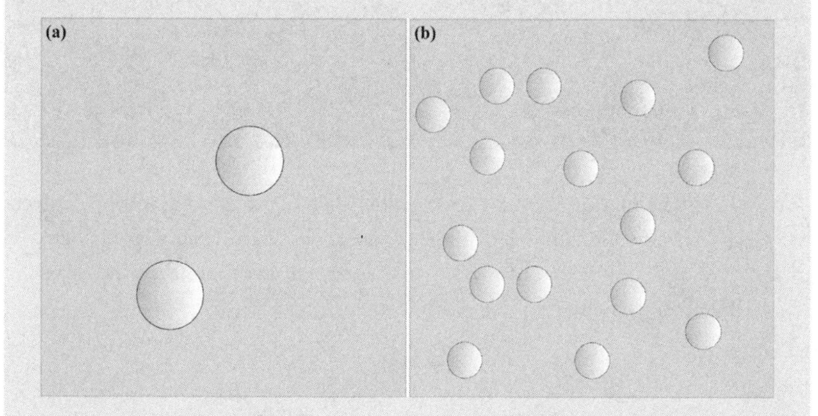

Box Figure 7.1 Airborne water droplets influence visibility. (*a*) Two droplets partially block our view of a gray square. (*b*) The same volume divided into eight smaller droplets blocks twice the area. Clearly, we will be able to see less far through air that contains small droplets than through air with the same volume of large droplets.

Figure 7.10 The marine layer over San Francisco Bay is a good example of advection fog. Here the Golden Gate Bridge is visible above the layer. (From "San Francisco Fog," Wikipedia, https://en.wikipedia.org/wiki/San_Francisco_fog#/media/File:San_francisco_in_fog_with_rays.jpg)

and surface. *Upslope fog* forms when moist air is blown up a hill (leading to cooling, and so on). *Evaporation fog* forms when cold air passes over warm water or wet land and often occurs in the fall. Note that the temperatures are reversed compared with advection fog: the air is cooler than the water. Fog requires a temperature difference between surface and the air immediately above it—which is warmer does not matter much.[10]

Precipitation

The 500,000 cubic kilometers (121,000 cubic miles) of water that falls on the surface of our planet every year does so mostly in the form of raindrops—some 10,000,000,000,000,000,000,000 of them, for those readers who are counting.[11] The solid forms of precipitation are much less numerous but are no less interesting. Here are two snippets to whet your appetite: snowflakes can take an hour to fall; hailstones can fall and rise several times before finally reaching the ground.

First, after the fashion of true meteorologists, let us list and briefly describe the seven types of precipitation that occur:

1. *Rain* is liquid water drops with a radius that exceeds 0.5 millimeter (0.02 inch). (The radius can be as much as 2.5 millimeters [0.1 inch].) Raindrops usually fall from nimbostratus or cumulonimbus clouds.
2. *Drizzle* is liquid water drops with a radius not exceeding 0.5 millimeter and usually falls from stratus clouds.
3. *Snow* is aggregates of ice crystals that fall from cumulus, cumulonimbus, nimbostratus, or altostratus clouds.
4. *Snow grains* are frozen drizzle.
5. *Sleet* is frozen raindrops with a radius of less than 2.5 millimeters.
6. *Hail* is frozen raindrops with a radius greater than 2.5 millimeters—sometimes much greater. Hail forms only in strong thunderstorms.
7. *Graupel* is snow pellets with a radius between 0.5 and 2.5 millimeters, formed when supercooled water droplets freeze onto a falling snowflake.

The rate at which rain falls is very variable in several different senses. First and most obviously, rainfall rate varies markedly from one region to another. The Atacama Desert of South America receives an average 15 millimeters (0.6 inch) of rain per year, whereas the village of Mawsynram in northeastern India receives an average of 11,871 millimeters (39 feet) of rain per year—most everywhere else is between these extremes.[12] The wettest place in the United States, and the eighth wettest in the world, is Mount Waialeale on the island of Kauai, Hawaii (figure 7.11).

Figure 7.11 View from a helicopter of Mount Waialeale, Kauai, Hawaii. (Photo by the author, March 2011)

At a given point on Earth—say, your back garden—the rate at which rain falls depends on many meteorological factors. Light rain corresponds to a rate of 2 to 4 millimeters (0.08 to 0.16 inch) per hour; a rate of 5 to 9 millimeters (0.20 to 0.35 inch) per hour is considered to be moderate rain; heavy rain is 10 to 40 millimeters (0.40 to 1.57 inches) per hour; violent rain is anything heavier than heavy. Rainfall rate varies in another sense: the speed at which raindrops descend to the ground depends on the updraft speed and on the diameter of the raindrops. Two forces control the speed of descent: gravity pulling down and aerodynamic drag pulling up. The drag force is complicated—it takes a different form for very small droplets than for larger ones, and the shape of large drops gets distorted with increasing speed (changing from spheres to jellyfish-like), which alters the aerodynamic drag. Empirically, we find from recent data that the speed at which a raindrop falls is proportional to its diameter. This means that larger drops overtake smaller ones, absorbing some smaller drops on their journey, and so their descent rate increases (assuming constant updraft speed).[13]

The awkward word *hydrometeor* is used technically to describe aqueous precipitation; it covers snow, hail, and so on, as well as raindrops. The solid forms have very different descent dynamics—different from one another and from raindrops. Hailstones are frozen rain but can grow larger than any raindrop. How so? As the hailstone falls through the air, it can pick up supercooled water droplets that freeze to the surface so the hailstone grows. Growth makes the stone heavier, and so it falls more easily. Sometimes, however, the updraft in a thunderstorm (thunderstorms are the subject of the next section) is strong enough to overcome the downward pull of gravity, and so the hailstone can grow quite large while suspended in the air. (In the appendix, we see how the size of a suspended hailstone depends on the square of the vertical wind speed.) Often the updraft speed varies so that a falling hailstone can be hit by a vertical gust that sweeps it back up to higher altitudes, where it accretes a new layer of ice, before becoming heavy enough to fall again. We know that this occurs because large hailstones, cut in half, show a layered structure like tree rings, indicating that the stone has yo-yoed up and down several times before striking the ground.[14]

Hailstorms occur mostly in the late afternoon and last for a median duration of six minutes. During that brief interval, however, a severe hailstorm can do a great deal of damage. Crops and livestock are often killed (200 sheep were killed in Montana in 1978, after being struck by baseball-size hailstones) and automobiles pummeled and smashed into write-offs.

The most common form of solid precipitation is altogether fluffier and more appealing. Snow is also more mysterious from the viewpoint of a physicist. A meteorologist considers it complicated for different reasons. There are steady snowfalls, and there are also flurries, squall lines, and blizzards. Snow on the ground compacts, melts, and refreezes. In short, snow on the ground can change its form with the changing surface temperature. It can feel powdery or wet and clinging. Its human appeal changes with time and temperature (and quantity; some places in Japan get 15.25 meters [50 feet] of snow per year, which must limit its appeal).

The process of snowflake formation is still largely a mystery, though recently physicists have made some theoretical progress. We understand why there is a basic hexagonal structure—due to the shape of water molecules—but not how these molecules get together around an ice nucleus in a supercooled cloud to form a snowflake. The folklore that no two flakes are alike is difficult to prove, but highly probable. What we do know about the formation of ice crystals (the aggregation of which we call snowflakes) is that it is strongly dependent on temperature. Thus if the air temperature is between 0°C (32°F) and –4°C (25°F), then ice crystals form as thin hexagonal plates. Between –4° and –6°C, they grow as needles; between –6° and –10°C, they develop as hexagonal columns. If the temperature is in the range –10° to –12°C, then we are back to plates; between –12° and –16°C, the crystals develop as dendrites or plates; between –16° and –22°C, plates dominate once again; while ice crystals grow as hollow needles when the temperature is in the range –22° to –40°C. Given that air temperature changes as a snowflake descends, then we expect its form to change as it grows.[15]

The change in air temperature with altitude—the temperature profile—can change the nature of precipitation as it falls so that what starts out as snow may land as sleet or rain, as suggested in figure 7.12.

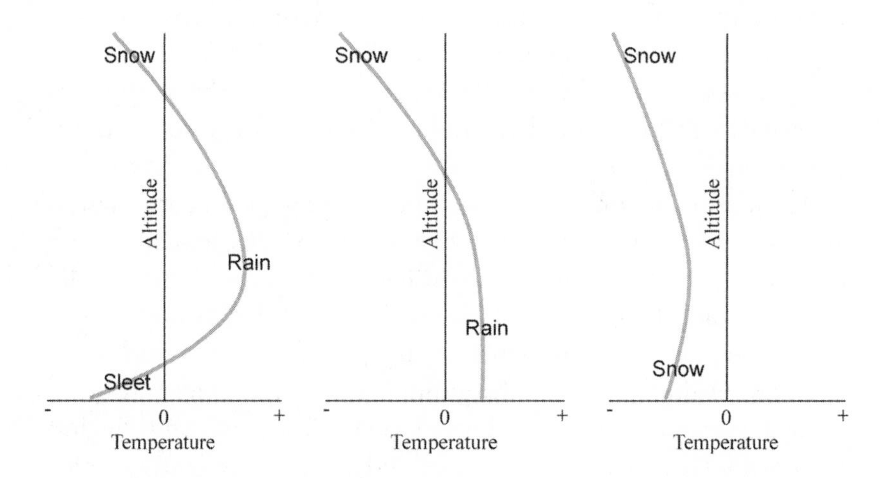

Figure 7.12 The influence of temperature profile on surface precipitation. In all three cases, the precipitation starts out aloft as snow, but it lands in different forms.

Thunderstorms

Thunder ("God rolling barrels across the loft floor," as my father used to say) results from lightning, as we all know. But how does lightning come about? Thunderstorms are rather extreme meteorological events in some ways, and in this section we have our first glimpse at them. We begin with the simplest and least violent type—the *single-cell* thunderstorm—and will then introduce you to the monstrous supercell variety.

A single-cell thunderstorm starts out as a strong, convective vertical air movement due to surface heating. It forms when moist air rises amid unstable air (air stability is unpacked when we examine the mechanisms involved in weather physics, in chapter 8). The key feature is that the air rises sufficiently high to reach regions of the upper troposphere where water droplets form and freeze; from the foregoing summary of cloud types, you can see that this considerable vertical extent leads to development of cumulonimbus clouds. Turbulence in the updraft causes ice particles to collide and ionize; the charged particles separate so that some parts of the sky build up positive electric charge, while others build up negative charge. Lightning, and so thunder, are the consequence when the charges become too great for the intervening atmosphere to keep apart.

There is little or no wind shear acting within a single-cell thunderstorm, and this feature limits the duration and ferocity of the storm; in fact, single-cell thunderstorms are self-extinguishing. Here is how that works. Moisture condenses in the rising air and develops ice particles that grow and then fall as precipitation, melting in the warmer air nearer the surface and landing as heavy rain—heavy because of the large volume of water vapor that is being rapidly converted into ice at the top of the thundercloud. The falling rain creates a downdraft of air that soon cancels the convective updraft that caused it. Hence single-cell thunderstorms are brief—typically lasting only 20 minutes. They are characterized by only one region of updraft, at the top of which (near the tropopause) the moist air branches out into a flattened layer so that the thundercloud looks like a mushroom (figure 7.13). Had much horizontal wind shear been present then this mushroom would have distorted into an anvil, characteristic of multicell or supercell thunderstorms, which are altogether more complex and powerful beasts.

Figure 7.13 Single-cell thunderstorm, characterized by a spreading cloud top that is more or less symmetrical, like a mushroom cap, in Arkansas, August 2014. This occurs because there is no wind shear to blow the top into an anvil shape, typical of more severe thunderstorms. (Photo by Griffinstorm, from "Thunderstorm," Wikipedia, https://en.wikipedia.org/wiki/Thunderstorm#/media/File:Single-cell_Thunderstorm_in_a_No-shear_Environment.jpg)

It's not that single-cell thunderstorms are weak or trivial events—they are not a mere puff of wind and sprinkle of water. Measurements of the precipitation from such storms permits a calculation of the energy consumed: a typical single-cell storm uses up around 10^{15} joules of energy (280 gigawatt hours), or about the energy of a tropical cyclone or of a small (240-kiloton TNT-equivalent) nuclear bomb. Worldwide, there are some 16 million thunderstorms of all types occurring annually, and some 2,000 at any given instant. Each drops about 500,000 tons of rainwater. In the United States, there are 100,000 thunderstorms per year, of which about 10% are severe. The brief lives of all single-cell thunderstorms follow the same pattern of birth, growth, and death. First is the developing stage, when cumulus clouds build up vertically, growing into tall towers. During this phase, there may be some lightning but no rain. The updraft continues during the mature stage of the storm, during which heavy rain falls with the accompanying downdraft. When this downdraft air strikes the surface, it splays outward radially, forming strong local winds known as the *gust front*. The storm is giving us everything it has got at this stage: strong winds, rain, hail, lightning, and thunder. Finally, during the dissipating phase, the downdraft dominates the updraft and so cuts off the source of the storm. The surface heat and the moist air at low altitudes can no longer reach the upper troposphere, and the storm dies. The gust front moves out and cuts off the supply of incoming air, which was rushing in from outside the storm area to fill the gap left by air that rose from the warm surface to start the storm.

It is important to understand that the limiting feature of a single-cell thunderstorm is the lack of wind shear. Had wind shear been present, then the downdraft region would not coincide with the updraft region and so would not cancel it. Thus thunderstorms that take place in air with significant horizontal shear do not self-extinguish: they last much longer and so can grow much larger.[16]

Thunderstorms are classed by their spatial extent and severity. We have looked at single-cell storms, which result from a single updraft region in air with no shear. *Multicell* storms are clusters of single-cell storms—that is, of individual updraft/downdraft regions. They last for a few hours, because the different component cells are at different stages of development, so the aggregate persists for longer than the individual component cells. A *squall line* is a linear series of severe storms that forms along or ahead of a cold front. A *supercell* thunderstorm is the most powerful type. It is a large and very severe storm that lasts for several hours and is characterized by separate updraft and downdraft regions due to shear, as explained. Most tornadoes spin off from supercell storms (box 7.2).

Lightning is an interesting phenomenon, one of nature's most mysterious, and so we will investigate it in a little more detail. We have seen that to develop a thunderstorm—that is, a lightning storm—the clouds have to be thick (3 to 4 kilometers [1.9 to 2.5 miles]) and cold, with a strong updraft. The charge separation in thunderclouds is as follows: the cold (–40°) tops are positively charged, the cool (–15°C [5°F]) middle section is negative, and the bottom parts are a mixture of positive and negative regions—necessarily separated from each other by insulating, electrically neutral air. The ground immediately beneath the bottom parts of the cloud is of opposite charge to the nearest cloud bottom part.

Box 7.2
Supercells

We have seen how wind shear influences the severity and duration of a thunderstorm: with no shear, precipitation falls through the updraft region and so quenches the updraft. There is no such quenching when shear is present because it causes the rain to fall over an area separate from the updraft. In addition, the cool gust front (from the downdraft) does not impede inflow of warm, moist air at the base. A supercell is a variety of wind-shear thunderstorm with a twist—quite literally.* The wind changes direction, as well as speed, with increasing altitude. This spiraling characteristic makes supercell storms more long lived than ordinary thunderstorms: several hours, rather than 30 minutes to an hour. Usually in the Northern Hemisphere the wind *veers*—changes direction clockwise as altitude increases.† Such veering winds introduce a cyclonic rotation to the updraft. A large, persistent updraft at the heart of a thunderstorm is called a *mesocyclone*, and it raises the thunderstorm to a new level—in fact, to the stratosphere. The structure of a supercell thunderstorm is distinct and reasonably well understood, and is illustrated in box figure 7.2.‡

Updraft wind speed at the heart of a mesocyclone can exceed 160 kilometers (100 miles) per hour compared with 48 kilometers (30 miles) per hour for an ordinary thunderstorm, and the air rises right to the top of the troposphere (the *tropopause*, separating troposphere and stratosphere). It naturally spreads out at this level because the stratospheric lapse rate is negative (that is, temperature increases with altitude [we investigate lapse rates in chapter 8]), and so this region is very stable. Air rises so fast in the updraft, however, that it overshoots the tropopause and penetrates a little into the stratosphere before settling down—this is the overshoot region shown in box figure 7.2. Wind shear creates an asymmetric spread of clouds at the tropopause, leading to the classic anvil structure of the cumulonimbus and mammatus clouds. Note the distinct updraft and downdraft regions. Note also the progress of a supercell thunderstorm from the point of view of someone in its path. First a shelf cloud appears, then light rain, then heavy rain. Hail commonly follows. Perhaps then there are strong winds developing into a tornado, or maybe the circulation remains in the wall cloud and does not touch ground. The rain is cool because it comes from high up and drags cool air down with it. It is commonly thought that hail is a harbinger of tornadoes, and as you can see it will often precede these twisters. Hail, however, is a more reliable indicator of the updraft strength in the mesocyclone. This is because hailstones that emerge from supercell storms can be very large (a couple of inches in diameter, very damaging to vehicles), and,

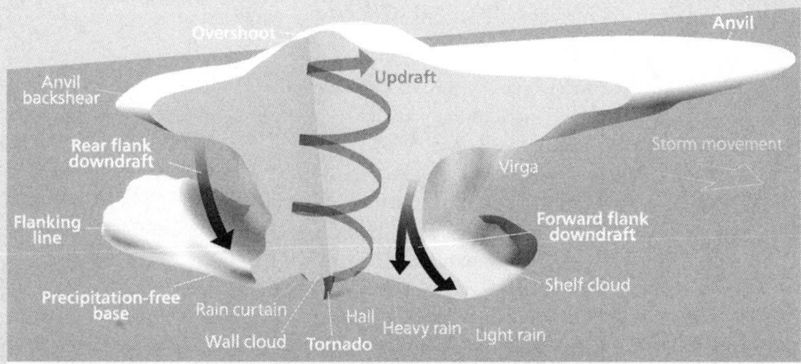

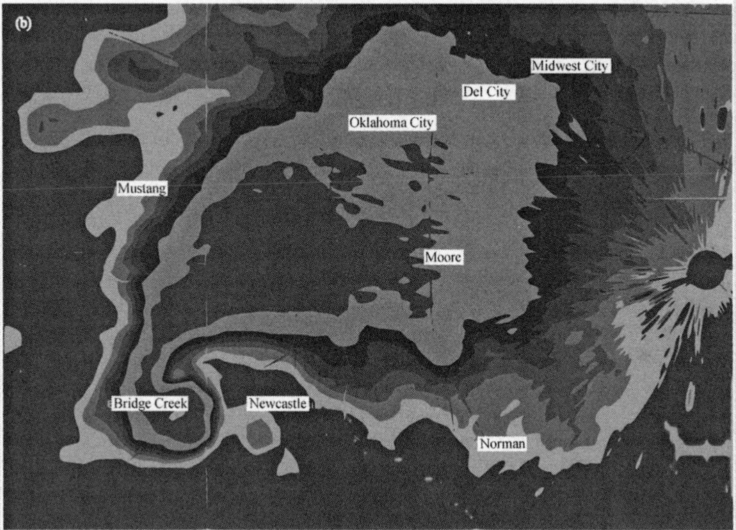

Box Figure 7.2 Classic Northern Hemisphere supercell thunderstorm. (*a*) Structure, viewed from the northwest. Note the small size of the tornado (which does not always develop) compared with the supercell. (*b*) Supercell viewed from above. This image is adapted from a radar view, at 6:57 P.M., of the supercell storm that ravaged central Oklahoma on May 3, 1999. The storm killed 40 people and injured 675. Radar location is the *dark circle, right center*. A powerful tornado is at the center of the "hook," located just by Bridge Creek. ([*a*] Illustration by Kelvin Song; [b] adapted from National Weather Service, "The Great Plains Tornado Outbreak of May 3-4, 1999," National Oceanic and Atmospheric Administration, http://www.srh.noaa.gov/oun/?n=events-19990503)

as we have seen, this means they were held up in the air for some time, as they build up size before falling. But big hailstones are heavy and so require a strong updraft to prevent them falling.[§]

[*]There are many online accounts of supercell formation and development. See, for example, National Oceanic and Atmospheric Administration, "Supercells," http://www.spc.noaa.gov/misc/AbtDerechos/supercells.htm; National Weather Service, National Oceanic and Atmospheric Administration, "Types of Thunderstorms," http://www.srh.noaa.gov/jetstream/tstorms/tstrmtypes.html; and "Supercell," Wikipedia, https://en.wikipedia.org/wiki/Supercell. For the role of vertical wind structure, see H. E. Brooks, C. A. Doswell III, and R. B. Wilhelmson, "The Role of Midtropospheric Winds in the Evolution and Maintenance of Low-Level Mesocyclones," *Monthly Weather Review* 122 (1994): 126–136. For historical data in Oklahoma, see J. E. Hocker and J. B. Basara, "A Geographical Information Systems-Based Analysis of Supercells Across Oklahoma from 1994 to 2003," *Journal of Applied Meteorology and Climatology* 47 (2007): 1518–1538. For impressive time-lapse videos of supercell formation, see "Wright to Newcastle, WY, Supercell Time-Lapse," May 18, 2014, https://m.youtube.com/watch?v=VoO89cqDgJU; and "Booker, Texas, Supercell Time Lapse," June 3, 2013, https://m.youtube.com/watch?v=mSORpd9QFSA.

[†]If direction changes counterclockwise as height increases, then the wind is called a *backing* wind. Veering winds are more common than backing winds in the Northern Hemisphere, most likely because of the Coriolis force: wind speed increases with altitude and the Coriolis force increases with speed, so the Coriolis force increases with altitude. Thus shearing winds will swerve more to the right—that is, clockwise—as height increases.

[‡]As with much of meteorology, you will by now be well aware, there is variability in supercells. There are at least three recognized types: low precipitation, high precipitation, and classic. Box figure 7.2 is of a classic supercell. Water vapor levels in the air are responsible for the different types, unsurprisingly. Low-precipitation supercells are the most common type in the lee of the Rockies (recall the Chinook effect and rain shadows), classic supercells predominate in the Great Plains, and high-precipitation supercells are the most common type farther east.

[§]A golfball-size hailstone requires an updraft speed of 90 kilometers (56 miles) per hour to prevent it from falling; a baseball-size hailstone needs an updraft of 160 kilometers per hour. Recall the calculation of hailstone size in the appendix. For technically minded readers, we note that gravitational pull on a spherical ball of ice (often supercell hailstones are not spherical—they are twisted into all sorts of fantastic shapes by the time they reach ground level—but it is reasonable to assume that they start out spherical) increases as the cube of its radius, whereas the drag force due to updraft increases as the square of both radius and updraft wind speed. This means that the updraft wind speed must increase as the square root of radius for the updraft to hold an ice ball aloft. As more and more water freezes to the ice ball surface, its radius increases to the point where the updraft can no longer support it, so it falls as hail. Thus large hailstones are an indication of fast updrafts.

As charge accumulates in these regions, the insulating capacity of surrounding air breaks down and lightning passes between regions of opposite charge, neutralizing the charge difference. The lightning is a plasma stream heated to some 50,000 K and is seen as brilliant white (usually) light. The most common type of discharge in thunderstorms is *in-cloud lightning*, sometimes called *sheet lightning* because it is seen as an undifferentiated flash with no linear or forked track. In-cloud lightning is just that—flashes within a single cloud—and accounts for one-quarter of all lightning. *Cloud-to-cloud* lightning is much rarer. *Cloud-to-ground* lightning is the most dangerous to humans for obvious reasons. *Ball lightning* is very rare and is not at all well understood (figure 7.14). There are several theories that can account for it, but none of these is yet widely accepted. It is quite small-scale compared with the others, which can cover hundreds of meters or several kilometers: the lightning balls are usually less than 50 centimeters (20 inches) across.[17]

A typical lightning flash carries about 30,000 amperes of current and transfers some 500 megajoules of energy. The energy transfer occurs over an interval of about a quarter of a second (in a complicated sequence of three or four separate strokes), and the peak power of a lightning bolt is tremendous: around 1 trillion watts. This is about one-twentieth of the electrical power consumption of the whole of humankind, albeit for only a few microseconds. This level of power and electrical current points to a voltage difference (of the two electrically

Figure 7.14 Ball lightning. This rare form of lightning lit up the sky over Maastricht, Netherlands, on June 28, 2011 (Photo by Joe Thomissen)

charged regions between which the lightning travels) of some tens of millions of volts. One particular set of thunderstorm voltage measurements reported negative values as low as –23 million volts relative to Earth, and positive values as high as +79 million volts.[18]

The physical shape (the branching structure and path shape) and development of individual lightning bolts are difficult to measure and are not well understood. The global distribution of lightning is much easier to ascertain. Seventy percent of lightning strikes are in the tropics. Globally, there are some 40 to 50 lightning flashes every second, on average, which adds up to about 1.5 billion a year. One aspect of lightning that is not widely appreciated outside the technical community is the role it plays in maintaining an electric field that naturally arises near the surface of Earth. Cosmic rays cause ionization in (where else?) the ionosphere, giving rise to an electric field of some 100 volts per meter between ionosphere and surface. This field would quickly discharge were it not for the action of thunderstorms across the globe, which act to maintain the voltage difference by pumping negative charge down to the surface.[19]

***

Clouds reflect incoming solar radiation and absorb outgoing longwave radiation significantly, but to varying degrees that depend on cloud type and altitude. They are a crucial component of Earth's energy budget, though one that is difficult to build into climate models because the physics operates on both small and large scales. Clouds are classified into 10 basic types, which aids meteorologists in assessing current and future atmospheric conditions. Fog is ground-level cloud that arises from the evaporation and condensation of local water. Precipitation is a major aspect of weather; the movement of water vapor and its condensation and precipitation transfer a large amount of energy, as exemplified dramatically by thunderstorms.

8

Weather Mechanisms

From where we stand the rain seems random. If we could stand somewhere else, we would see the order in it.

Tony Hillerman

In chapter 4, we examined the slow dynamical effects that drive climates and climate change; here we investigate the faster dynamics (hours, rather than eons) that drive weather and weather change. In this chapter, I draw together many of the threads that arose in earlier chapters—summarize the story so far—and then weave them into a fabric that permits us to see the whole tapestry. Some physical forces plus a few simple physical laws give rise to the immensely complex phenomenon of weather. We will see how well-known weather phenomena arise from these basic physical principles. A detailed development that takes us from basic laws to a quantitative understanding of weather phenomena involves much complex mathematics, but we will not take that circuitous route. Qualitative insight will follow by connecting the dots directly with a math-free approach if you are prepared to accept certain assertions without proof—they would require a university physics degree. By forgoing the degree and accepting this intuitive assertion, and a few others, we will get where we want to go.[1]

They say that when speaking in public, you should follow three steps: tell them what you are going to tell them, then tell them, then tell them what you have told them. By so doing, you infuse your audience with the content of your speech. You may have noticed that, in this book, I have been doing something similar. But each retelling is not quite a repetition, because the emphasis is different and more content is added each time. I make no apologies for this approach—which continues in this chapter—I find it to be the best way to convey complex ideas.

The Story So Far

Our planet's weather system is driven by electromagnetic (EM) energy from the sun. We saw in chapters 1 and 2 that incoming shortwave solar radiation heats up the atmosphere a little, and heats it up much more as outgoing longwave blackbody radiation after absorption by Earth's surface. Atmospheric conditions vary because our planet rotates. Thus the density of EM radiation that reaches the top of the atmosphere is pretty constant, hour to hour throughout the year, but varies cyclically at the surface with a 24-hour period. Further variation arises at the surface because of cloud cover, which changes the amount of radiation that is absorbed or reflected back into space, as we saw in chapter 7. Yet more atmospheric variation arises from differential heating: equatorial regions and most land masses get hotter than polar regions and oceans, because they receive more solar radiation or absorb more radiation. Thus different latitudes and longitudes on the surface of Earth have different temperatures.

I adopt the convention, just for this chapter, of italicizing the assertions that we will need without proof to help us get to an understanding of fundamental weather mechanisms and phenomena. These assertions will be simple and intuitive, so you won't be wondering "where the heck did he get *that* from?" Here we go: *hot air rises*. Thus a location on Earth's surface that is warmer than surrounding regions will cause the air above it to rise. But *nature abhors a vacuum*, and so the surface air that rises must be replaced by air rushing in from the sides—we call such air "winds." (A ubiquitous example of wind created by differential heating, familiar to readers who live along the coast, is a *sea breeze* [box 8.1].) So we see how a rotating Earth that is being heated by the sun will develop both vertical and horizontal movement of its atmosphere. This movement is the foundation of weather.

Box 8.1
Sea Breeze

A sea breeze, or onshore wind, arises typically on a summer afternoon following several hours of sunshine. The *sea has a greater heat capacity than the land*, which means that it can absorb more heat before it begins to rise in temperature. In particular, if the sea and adjacent land absorb the same amount of sunshine (that is, solar heat), then the land will heat up more. So by afternoon, the air is rising above the land. The rising air leaves a partial vacuum beneath it that is filled in by cooler air from above the surface of the sea. Put another way, rising air creates a lower pressure over the land surface compared with pressure over the sea surface, and the pressure difference pushes air from sea to land. The rising air cools and falls back to the surface over the sea, so that a circulation pattern of air flow is set up, powered by the sun. The result for a person standing on the shore is a cool breeze coming in from the sea.

At night, the land cools and can become cooler than the sea. If this happens, then the circulation pattern reverses: air flows from land to sea, near the surface, pushing air up over the sea; this rising air cools and flows back down to the surface over the land. We call the resulting wind a *land breeze*.

Due to evaporation from the surface of oceans (mostly) and to transpiration from the surface of tree leaves (to a lesser extent, though important locally), there is a vast amount of water vapor in the atmosphere, as we saw in chapter 7. The water vapor content of air is limited by temperature, and *the maximum water vapor content increases as temperature increases*, as shown in figure 8.1. If the rising air contains moisture, then at some altitude the temperature will drop enough so that the air can no longer contain all the moisture within it (this situation corresponds, on the graph of figure 8.1, to air moving from right to left across the line). So the moisture condenses to form clouds. This process is dynamical and complex, and we can easily see why. First, the altitude at which moisture condenses depends on the temperature profile—the change in temperature with increasing altitude—called the *lapse rate* by meteorologists. Lapse rate is a sufficiently important subject to merit its own section, later in the chapter. Second, as we saw in chapter 3, pressure falls exponentially with increasing altitude, so that rising air expands, thus reducing its density and therefore, according to the *First Law of Thermodynamics*,[2] its temperature.

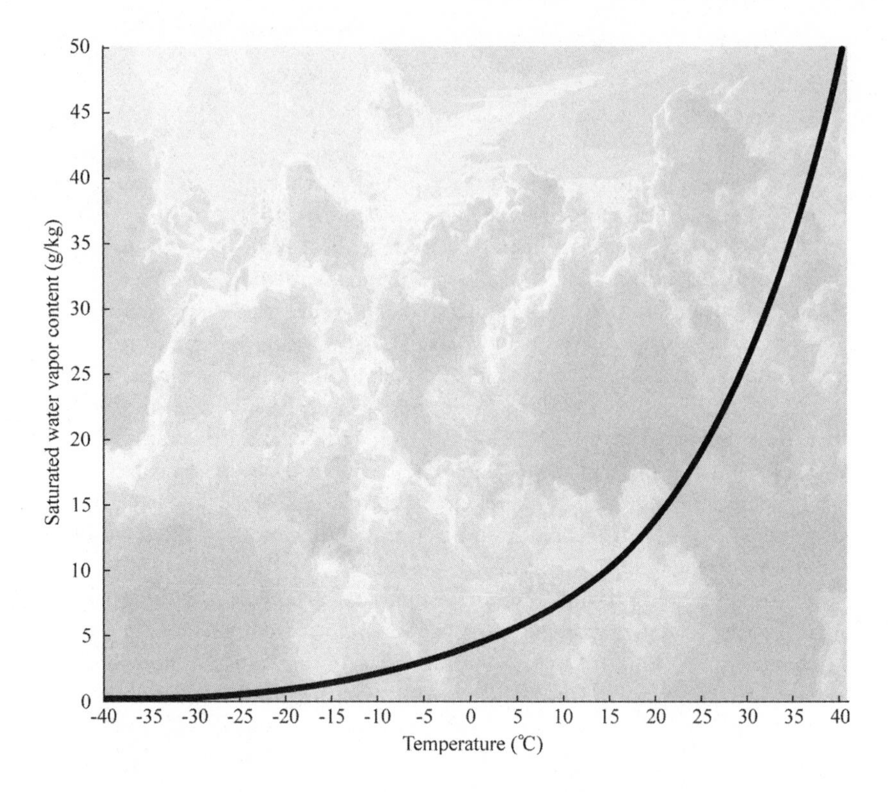

Figure 8.1 Maximum—that is, saturated—water-vapor content of air versus air temperature, for sea-level barometric pressure. Thus, for example, warm air at 30°C (86°F) can hold about five times as much water as cold air at 0°C (32°F).

Third, water condensing from vapor to liquid gives up heat (this is simply the reverse process of heating water to cause it to boil), thus changing the air temperature. So we see that there is a stew of interacting variables that influence what happens when a parcel of moist air rises.[3] To make accurate predictions, meteorologists have to know

- The initial temperature and pressure of the air
- The water content of the air
- The lapse rate

Of course, they also must understand the physics of the process. We will see later in the chapter just how divergent the behaviors of a parcel of moist air can be. Clearly, it is important to get these predictions right because rising moist air leads to cloud formation and movement (possibly), precipitation (perhaps), weak or strong winds, temperature change at the surface (maybe)—in short, much of our weather.

Heat from the sun, including that which is reradiated from Earth, causes water to evaporate and air to rise.[4] Rising air causes winds and clouds and heat transfer. These are the weather mechanisms that we have seen so far. Now we turn to a summary of the basic physical forces that give rise to these, and other, mechanisms that drive our weather.

A Stampede of Forces

Some areas of research in the more esoteric branches of physics are complicated and hard to understand because the forces involved are complicated and hard to understand. Thus researchers do not understand how gravity works at the center of a black hole. Researchers understand the forces that arise from quantum mechanics but find them hard to believe and to interpret. But atmospheric physics in general and meteorology in particular are not complicated and hard to understand for this reason: the forces that drive our atmosphere are everyday forces that we humans can get to grips with intuitively—I list them here and expand on each a little. What makes meteorology so very challenging is the many, many variables that feed into it from different branches of physics, chemistry, and biology. The resulting complexity is that which arises from simple forces acting on many parts, rather than the complexity of complicated forces acting on a few parts. This is why meteorological prediction requires vast amounts of accurate input data and knowledge of their statistics, as we saw in chapters 5 and 6, and why it needs enormous number-crunching computing facilities, specially built and manned by armies of trained people to predict the weather a few days into the future, as we will see in chapter 10.

The forces of nature that drive our weather are

- Gravity
- Buoyancy
- Pressure gradients
- Friction
- Coriolis force

There are other, minor influences, such as the centrifugal force that results from Earth's rotation and seismic forces that occasionally spew vast amounts of dust into the atmosphere. To keep this book a reasonable length, I have to draw the line somewhere, and these extra forces are on the other side of the line. They and others are interesting but not crucial and so are not a part of the story that I have to tell.

Gravity is so obvious (though it took an Isaac Newton to make it so) that I can get it across to you in a couple of sentences. Gravity is a force that is directed toward the center of Earth. There are smaller components directed toward the centers of the sun and moon (and much smaller components directed toward Jupiter, Alpha Centauri, Andromeda . . .) that, because our planet rotates, are responsible for the tides. Gravity is the force of attraction between masses and is approximately constant over the surface of Earth. Here we accept it as a constant, downward-acting force and move on to more interesting forces.

Buoyancy is reduced gravity due to differences in density, and so acts vertically. It is a familiar force to anyone who has tried to push a beach ball underwater. It is clearly very important to weather physics in the context of air packets rising or falling based on temperature difference (and hence density difference).

Pressure is force per unit area. You press down on measuring scales with your feet, and the force (due to gravity) is just your weight. The pressure under your feet is the force divided by the area of your feet. Similarly, the force pushing down on the top of your head from the column of air that is above you (all the way up into space, where the atmosphere becomes thinner and thinner until it disappears) is just the weight of the column of air. It exerts a pressure of 14.7 pounds per square inch at sea level. To calculate the force due to this pressure, multiply by the area of your head.

Rising parcels of air leave reduced pressure underneath them, less than that of the surrounding air, which rushes in to fill the void—propelled by the force that arises from the pressure difference. Buoyancy is a vertical force that results from density difference. That parcel of air rises because air around it is pulled down by gravity more strongly (it is denser, and so more massive, and so feels a greater gravitational pull). The parcel gets displaced—squeezed upward—and it rises until it reaches an altitude at which its density matches that of the surrounding air.[5]

Friction is a very important type of force in our everyday lives and it takes many forms. Here are two examples. If you live in a wooden house that is held together by nails, then you rely on friction to keep your house from falling apart. Lubrication in your car acts to reduce the friction between components that slide against each other; such friction results in wear and causes the components to heat up. The friction that applies to meteorology is of the second type. Imagine hauling a heavy trunk across your attic floor. The trunk and floor are in contact, and the force of friction between them resists the movement. Replace the floor by a landmass and the trunk by a wind blowing across it, and we have a similar sliding friction acting to retard the movement of the wind. (The significance of friction forces between ocean and atmosphere is brought home in figure 8.2.)

Figure 8.2 The force of friction. Here waves strike a sea wall during a storm in New England, 1938. The waves spurt upward, like geysers. They are driven against the sea wall by wind, via friction with the ocean surface. (From National Weather Service Collection, National Oceanic and Atmospheric Administration Photo Library, http://www.photolib.noaa.gov/htmls/wea00412.htm)

The scenarios are similar but not the same. Air is a fluid, unlike your trunk, and so it can *shear*. The result of friction between Earth's surface and a wind that blows over it is that the wind nearest the surface (the *boundary layer*) is slowed down to nothing, whereas the wind higher up (the *free flow* region) is slowed down less and less. The result is a wind profile; at 15.25 or 30.5 meters (50 or 100 feet) or 300 meters (1,000 feet) above the surface, the wind speed is, say, 20 miles per hour, but get closer to the ground and you find the wind speed reducing, to zero at ground level. More important for us is the effect of wind over the oceans. Seawater is also a fluid, and so it is moved along by the winds that act above it, in the same direction. The sea surface is moved by the force of friction between sea and air. Thus ocean currents arise. In strong winds, ocean waves are whipped up, especially when the wind has a big *reach*—when the area of ocean over which the wind acts is large. Tides may be due to gravity, but open-ocean waves (be they small ripples in a light breeze or monstrous wave trains in a storm) are due to friction between wind and water.

The Coriolis force, named after the nineteenth-century French scientist who first explained it, is the weirdest and least intuitive of the forces that interest us. For that reason, I explain it in more detail in box 8.2. Not the least reason for its weirdness is that its existence depends on your viewpoint. If you are on the surface of Earth and detect air movement that is of a certain type—a spiraling effect—then you might say "Aha! The Coriolis force is acting on this parcel of air!" whereas your friend in space might look down on the same parcel of air, way below him, and see no force acting at all. The Coriolis force, which we first met in chapter 2, is called a *fictitious force* by physicists for this reason (though "apparent force" might have been a better description). Its magnitude depends on your point of view—your frame of reference, a physicist would say.

Most humans are confined to the surface of Earth, so we rotate with our planet as it hurtles through space. The rotation gives rise to the Coriolis force, and in meteorology we must take it into account for large-scale weather phenomena, such as hurricanes. The weird effect of the Coriolis force is that it makes any object that is moving over the surface of Earth deviate to the right, if it is in the Northern Hemisphere, and to the left, if it is in the Southern Hemisphere. The force that causes the deviation depends on object speed as well as

Box 8.2
Coriolis Force

You are a spectator standing beside a flat-topped merry-go-round (box figure 8.2). Your friend Ollie is on board the merry-go-round, sitting at the edge. He releases a ball that you observe to travel in a straight line passing through the center of the merry-go-round and falling off the far edge. Ollie sees things very differently. He sees the ball move away to his left, and then curve right and leave the merry-go-round close by where he released it. The different opinions about the ball trajectory are both "correct," but seem to differ because of the different viewpoints. In particular, Ollie sees that a force must be acting on the ball to cause it to follow a curved trajectory. You see no force. For a counterclockwise-rotating merry-go-round, this Coriolis force causes a moving object to veer to the right, as in the figure. Now move up from two dimensions to three: the merry-go-round is replaced by Earth; you are a spectator in space directly above the North Pole, looking down. Ollie is on the surface of Earth. The ball is a balloon floating in the wind. *Voilà*—Ollie and the rest of us in the Northern Hemisphere see the balloon (the wind) curve to the right due to the Coriolis force.

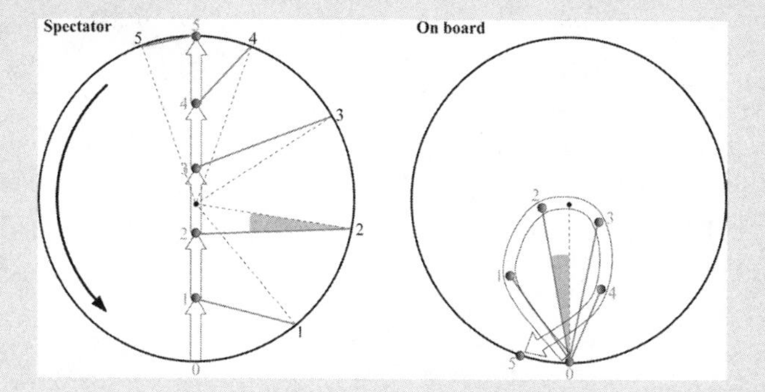

Box Figure 8.2 Coriolis force. (*Left*) You stand beside a merry-go-round as Ollie, on the merry-go-round at 0, releases a ball. You see the ball follow a straight line (*white arrows*). When Ollie is at positions 1, 2, 3, 4, 5 on the perimeter, the ball is at positions 1, 2, 3, 4, 5 on the straight line passing through the center. When Ollie is at position 2, you note the angle (*shaded area*) he sees between the ball and the center. (*Right*) Ollie sees the ball follow this curved trajectory (*white arrow*). Note that the angle at position 2 is the same as before; indeed, the angle for all points on the ball trajectory is the same from both viewpoints: Spectator and On board. The distance between Ollie and the ball, at any given time, is also the same in both frames. Ollie sees a force acting on the ball, causing it to curve. This is the Coriolis force; it is called fictitious because its magnitude depends on the frame of reference (point of view).

on object latitude and longitude. For small-scale movement, such as throwing a baseball from pitcher to catcher, the effect is completely negligible. Over the distance between pitcher and catcher, the Coriolis force will cause a baseball to turn to the right by a tiny fraction of an inch. But over 160 kilometers (100 miles), it becomes important. Long-range artillery gunlayers need to account for the Coriolis force if the shells they fire are to hit the intended target. Strong winds that sweep across large areas deviate to the right (in the Northern Hemisphere) and so go into a spiraling motion, as we see in figure 8.3. If the wind speed is due to a low-pressure area nearby, then the Coriolis

Figure 8.3 The Coriolis force in action. Cyclones are at the heart of massive tropical storms and of storms that form at mid-latitudes. Here is a pair of cyclones that formed between Iceland (*top center*) and Scotland (*lower right*) on November 20, 2006. At the center of each is a low-pressure area that pulls in air from all sides, toward the center; the Coriolis force causes this moving air to deviate to the right—hence the observed counterclockwise spirals. In the Southern Hemisphere, these spirals would be clockwise. (Image by Jesse Allen, NASA, Earth Observatory, http://earthobservatory.nasa.gov/IOTD/view.php?id=7264)

deviation changes the wind movement into a counterclockwise corkscrew that we call a *cyclone*. Cyclones and anticyclones are discussed further in a later section.

Re: Lapses

There are three lapse rates for air, and they depend on the thermodynamic properties of air, on its water vapor content, and on surface temperature.

1. The *environmental lapse rate* is that which applies to a stable atmosphere in equilibrium, meaning one that is not moving. Given that ground beneath the air is heated by the sun and that pressure decreases with altitude, we expect air temperature to decrease with increasing altitude. The average rate is 6.5°C (11.7°F) per kilometer (0.6 mile), but it varies with local surface temperature.
2. The *dry adiabatic lapse rate* is that at which a parcel of unsaturated air cools as it rises; this rate can be calculated from thermodynamics and is a constant for a given planet with a given atmosphere.[6] On Earth, it is 9.8°C (17.6°F) per kilometer increase in altitude. It would be greater on a bigger planet with the same type of atmosphere as Earth; it would be different (greater or smaller) on an Earth with a different composition of atmospheric gases. When the air is saturated with water vapor, then the lapse rate is different. This is because saturated air that is cooled loses its water vapor through condensation; condensation releases energy (latent heat) that was formerly hidden within the vapor, thus slowing the lapse rate.
3. The resultant *saturated adiabatic lapse rate* depends on temperature and is typically 5.5°C (9.9°F) to 6.5°C per kilometer near the surface. (In the United States, lapse rates are usually expressed in degrees Fahrenheit per 1,000 feet increase in altitude, so the dry adiabatic lapse rate is 5.5°F per 1,000 feet, and the average saturated adiabatic lapse rate is about 3°F per 1,000 feet.)

Lapse rates are dull, but important. A simple weather example of the significance of different lapse rates for saturated and unsaturated air is that of Chinook winds (box 8.3).

Box 8.3
Chinook

In meteorology, *föehn* is the word used to describe warm, dry winds that descend the lee slope of mountains. In western North America, such a wind is referred to as a *Chinook*, a word taken from the indigenous coastal people of British Columbia.* Here we see how such winds arise and what weather they instigate.

In box figure 8.3, we see a winter wind blowing from the southwest and approaching the Rocky Mountains—say, in Montana. The wind is saturated with water; weather forecasters in the Pacific Northwest call this wind a *Pineapple Express*, as it seems to come from Hawaii.† In this case, the wind blows over the western foothills at 40°F (4.4°C). What weather does it give rise to, farther east? As it rises up the mountain slopes, it cools at the saturated rate. It can no longer hold all the water, and so these western slopes experience rain. At the top of a 10,000-foot (3,000-meter) mountain, the wind has cooled to 10°F (–12.2°C). A lenticular cloud known as a *chinook arch* (or *föehn arch* or *helm cloud*) may form at the summit (see figure 7.8). The descending air is dry, and it warms at the dry lapse rate as it descends the lee slope.‡ When it reaches the mountain

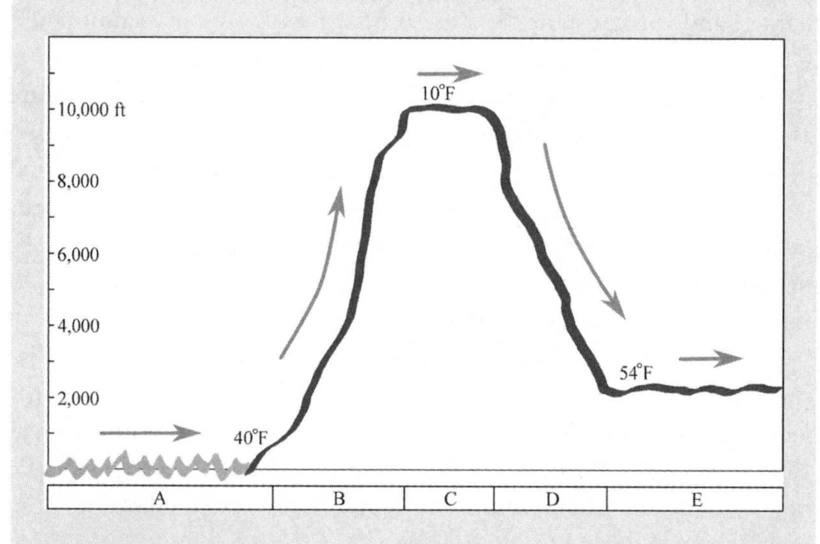

Box Figure 8.3 Warm, moist air at sea level is forced upward by a mountain; the air cools, and its water vapor condenses, resulting in rain on the windward slope. When the now-dry air descends the lee slope, it warms, creating a Chinook wind.

bottom, it is at 54°F (12.2°C) and is bone dry. Warm and dry—Chinook winds melt snow quickly.

Thus we are able to predict the weather in the five zones, A to E, shown in the figure, given the initial wind temperature, its water vapor content, and the topography. Also, please note for later, we are assuming that the atmosphere in the vicinity is stable. In zone A, the weather is windy and humid, at 40°F. Zone B is rainy, with air temperature falling at higher elevations, to 10°F at the summit, zone C, where it is cloudy. Zone D has a dry wind increasingly warm at lower elevations. In zone E, the air is desiccating and the temperature is 54°F. Note that at 2,000 feet (610 meters) elevation in winter, we might expect the ground temperature in zone E to be much lower, say –20°F (–28.9°C); it is the contrast with air temperature that makes the Chinook winds so remarkable.

*Or *Santa Ana*, for the warm downslope winds in southern California. There are also cold downslope winds, such as the *Mistral* of the Rhône Valley in France.

†Regular and prominent meteorological phenomena, particularly storms, are often assigned whimsical names. As well as the Pineapple Express, Americans feel the effects of the *Siberian Express* (Arctic weather in the central and eastern United States), the *Alberta Clipper* (strong, cold winds in the lee of the Rockies, well known in Chicago), the *Hatteras Bomb* (fierce blizzard off Cape Hatteras, North Carolina), the *Colorado Low* (the main source of winter storms in the central United States), and the *Chattanooga Choo Choo* (storm progressing slowly from the Appalachians to the northeastern United States).

‡Often the air that rises from lower elevations and passes over the summit is warmer than the air surrounding it; nevertheless, this warmer air is forced down the lee slope by hydrodynamic effects that overmatch the buoyant force caused by a difference in density.

There's Nothing Stable in the World; Uproar's Your Only Music

Contrary to Keats's line, the atmosphere can be stable—and it can be in uproar. Whether the atmosphere is stable or unstable depends on lapse rate—that is, on the rate at which temperature changes with increasing altitude. The connection between stability and lapse rate is not obvious, but the chain of reasoning is easy to grasp.

We have seen that differential heating of the atmosphere leads to pressure differences, which lead to horizontal movements of air. These horizontal movements are modified by the Coriolis force and can become spiraling cyclones (with low-pressure centers) or anticyclones (high pressure). Temperature difference also leads to vertical movement of air parcels, which give rise to circulation patterns such as those responsible for sea breezes. It is the vertical movement of air that holds the key to atmospheric stability and instability.

Now that we know the basics of temperature lapse rates, we can understand what meteorologists mean when they refer to "stable air" or "unstable atmospheric conditions." Suppose that a parcel of air rises; perhaps it is driven upward by a mountain range, as are Chinook winds. As the air rises it cools, at either the dry or the saturated lapse rate, depending on its water vapor content. Suppose that the environmental lapse rate in the vicinity of our parcel is less than the lapse rate of the parcel. This means that the surrounding air cools more slowly with increasing altitude than does our parcel. If the surrounding air is warmer, then it is less dense, so the parcel sinks back down. Similarly, if the parcel falls to lower altitudes (say, it descends into a valley),then it finds itself surrounded by cooler, denser air and so it rises again. Thus if the environmental lapse rate is less than the parcel lapse rate, the air is *stable* because it resists changes in vertical air movement. Any slight change in altitude by a parcel of air is countered by the surrounding air, and the parcel returns to its original position. Thus the atmosphere is layered and stays that way.

We saw earlier, in chapter 3, that the stratosphere is characterized by temperatures that rise with increasing altitude—that is, a negative environmental lapse rate.[7] This lapse rate is certainly less than the dry or the saturated adiabatic lapse rate of any parcel of air that finds itself in the stratosphere, and so the stratosphere is stable. Closer to the surface, stable air masses result in weather that changes only slowly; it is calm. Clear and calm nights, or days of continuous sun or of gentle rain or of persistent fog, indicate stable air conditions.

Now suppose that the environmental lapse rate near a parcel of rising air exceeds the dry or the saturated lapse rate of the parcel. In this case, as the parcel rises, it cools more slowly than the surrounding air and so ends up surrounded by denser air. So buoyancy forces the parcel upward faster—the initial upward movement of the parcel is reinforced. Similarly, if the parcel begins to sink then this movement is reinforced and the descent rate accelerates. Such air is *unstable*. It is unstable in exactly the same sense as a pencil balanced on its point is unstable: if perfectly vertical, it will stay so; but this equilibrium is unstable because the slightest movement will cause the pencil to fall.

The weather associated with unstable air is variable in both time and space: it changes quickly from one hour to the next and from one town to the next. Unstable air is turbulent; expect thunderstorms. Typically, the environmental lapse rate decreases at night as air near Earth's surface cools faster than air that is higher up. So it is often the case that the atmosphere in a given location—say, near your home town—can be stable at night and unstable during the day. Stable and unstable air is conveniently illustrated in figure 8.4.

A little thought shows that stable air is an example of negative feedback and of statistical equilibrium. Unstable air is an example of positive feedback and of sensitivity to initial conditions (small differences grow into large ones). I will let this point slide for a page or two, but please bear it in mind.

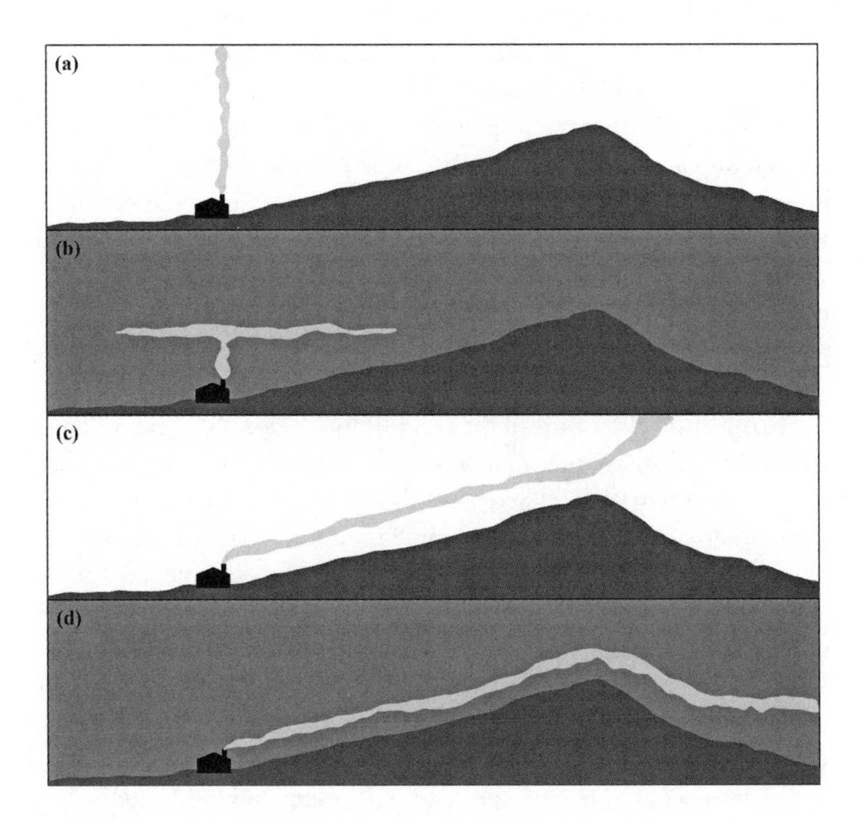

Figure 8.4 Stable and unstable air. Here the air is stable at night and unstable during the day—a common occurrence. (*a*) Day, no wind. (*b*) Night, no wind. (*c*) Day, wind. (*d*) Night, wind.

Most of the troposphere is neither stable nor unstable, most of the time. At any given location near the surface of Earth, two-thirds of the time the atmosphere is *conditionally stable*. This means that the environmental lapse rate is greater than the saturated adiabatic lapse rate but less than the dry adiabatic lapse rate. In this case, what happens to a parcel of air that is raised through the surrounding air depends on the parcel's water vapor content. If the parcel is saturated, then it acts as if the atmosphere is unstable, whereas if it is unsaturated, it acts as if the atmosphere is stable.

This complication requires a little thought and has consequences. If a parcel of air is unsaturated at the surface, it is stable and will not rise by thermal convection. If it is *forced* to rise—say, by being blown over a mountain—then it will sink back down to the surface. BUT the parcel cools as it rises and so at some critical altitude becomes saturated. Above the critical altitude (the *lifted condensation level*), therefore, any rise is reinforced as the saturated parcel is in unstable air. So conditional stability is characterized by a critical altitude below which the air is stable and above which it is unstable. If a mountain peak is below the lifted condensation level, then air that is forced up over the mountain will flow down the lee side; if the peak is above this level, then the air will continue to rise. (Note that the lifted condensation level is not the same as the *level of free convection*, which is the altitude at which a rising parcel of air first becomes warmer than its surroundings. These levels, and conditional stability, are illustrated in figure 8.5.)

We can appreciate some of the complications faced by meteorologists who must calculate quantitatively what happens to our atmosphere over the next few days. Here I have shown how one simple scenario—the Chinook winds—works, but my arguments were qualitative. The physics is correct, and the weather predictions (for example, rain on the windward slope) are right on, but meteorologists are able and expected to predict more. They will tell us the likelihood of rain, how much rain is expected at what locations, when the rain will occur, and what the temperature will be. Consider the information that they need to make such predictions: accurate knowledge of the water vapor content of the onshore wind, the wind speed and direction, and local topography, plus knowledge of the atmospheric stability of the region. Suppose, for example, that the wind approaching the coast (see box 8.3) carries a large amount of water vapor but is not saturated. The predictions of

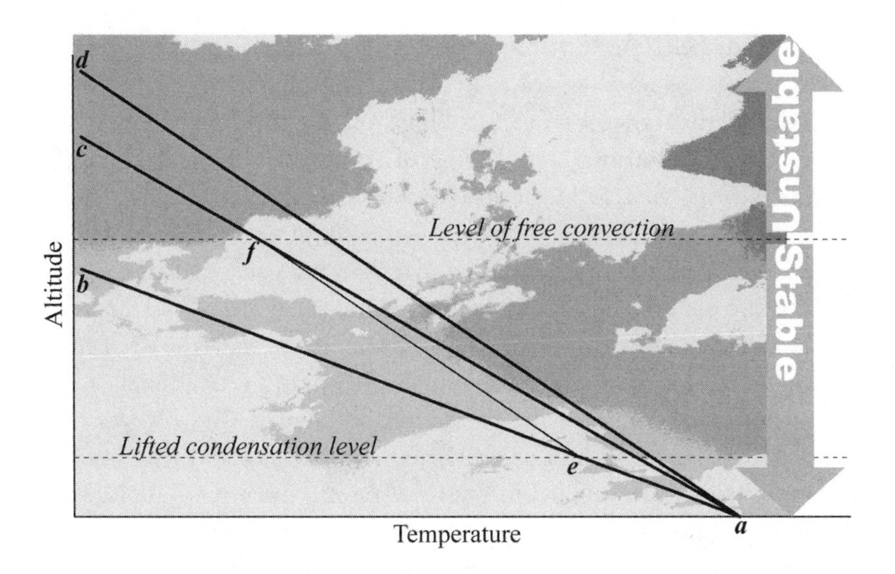

Figure 8.5 Conditional stability. The air below the level of free convection is stable and that above is unstable. Temperature versus altitude: dry adiabatic lapse rate *ab*, environmental lapse rate *ac*, saturated adiabatic lapse rate *ad*. An unsaturated parcel near the surface will follow the path *aef*: it rises at the dry lapse rate until its temperature drops to the dew point (the temperature at which water vapor condenses), at which it is saturated and thereafter rises at the saturated lapse rate. Clouds form. If the parcel becomes free to choose its own level before it reaches *f* (say, it passes over the top of a mountain), then it will descend to the surface. If it becomes free only when above *f*, then it will ascend until its temperature matches that of the surrounding air.

regional weather will differ from the saturated case. You can see why: the air is forced up the mountain and cools initially at the unsaturated lapse rate. When it reaches the condensation level (which meteorologists must calculate; you can see from figure 8.1 that it depends on temperature, which they must either calculate or measure), it begins to rain and the air continues up the mountain, cooling at the saturated lapse rate. Thus the onset of rain occurs higher up the mountain than for the saturated case, and the air temperatures all across the mountain will be different than for the saturated case. If the atmosphere in the vicinity of the mountain is conditionally stable instead of stable and the mountain height is below the level of free convection, there may be no Chinook winds on the lee side.

In a nutshell, the daytime atmosphere is usually unstable for the same reason that boiling water in a pot is unstable. It is heated from below and thus gives rise to a rolling boil (in the case of water) or a more sedate churning and mixing of vertical layers as heat rises by convection (in the case of the atmosphere). Sometimes the air is cooler near the surface than it is higher up—this situation is unusual, and is called a temperature inversion.[8] An inversion is a particular case of stable air in which the environmental lapse rate is negative—the air heats up as altitude increases—and so, of course, it is less than the dry or the saturated lapse rate of any parcel of air. The boiling-water analogy also serves to show how hard the prediction problem is. Yes, physics tells us that the water will boil when heated and bubbles will rise, but for weather prediction we need to know *where and when* the bubbles rise. Only acute knowledge of the initial conditions as well as of the relevant physics and statistics will give us any hope of such detailed predictions, for reasons we thrashed out in chapter 6.

Enter the Vortex

In addition to the forces that act on the atmosphere, several laws of physics influence our weather. These laws are useful for meteorologists when calculating weather evolution, as they are universal principles that apply everywhere at all times. Here I discuss two (there are many others, but these two are important enough in meteorology to be emphasized): the conservation of energy/First Law of Thermodynamics and the conservation of angular momentum. Let me explain what these laws mean and why they matter to us. Then we will be in a position to understand better vortex phenomena such as typhoons and tornadoes.

We have already encountered energy conservation. Consider a parcel of air at the surface. During the day, the surface is heated by the sun, and so the parcel is heated. This heat energy causes the parcel to expand and so do work. By expanding, it becomes less dense and so it rises, gaining gravitational potential energy. The difference between the input energy (heat) and the work done (rising, and displacing air) is manifest as changed internal energy (changed temperature). Energy conservation is simply a balancing of the energy books, which physicists have a very good handle on and can calculate accurately.[9]

It is important to note that energy conservation means that no energy is created or destroyed; this is not quite the same as saying that the system we are interested in (say, a parcel of air) has the same energy at all times. It could be that energy is dissipated (lost from the parcel) by friction or that heat leaks in or out, which may lead to the energy of the system changing in time. Only if the system is isolated can we say for sure that it conserves energy; if it is not isolated (after all, parcels of air are surrounded by other air), then energy can leak into the surroundings. However, it is often useful to approximate a system as energy conserving. For example, we saw earlier that parcels of air can be thought of as developing adiabatically, which is just a way of saying that they do not exchange heat with the environment.

The conservation of angular momentum is a geometrical law. It is more complicated than energy conservation because angular momentum is a vector quantity: it has a direction as well as a magnitude. Both magnitude and direction are conserved. Momentum is a measure of how hard it is to stop a moving object. A cannonball is harder to stop than a bullet moving at the same speed because it is more massive. A cannonball is harder to stop than a bowling ball of the same mass because it is moving faster. Angular momentum is the angular equivalent of momentum: it is a measure of how hard it is to stop a rotating object. Angular momentum depends on mass and angular speed, but also on size. Imagine a bicycle tire on a lubricated horizontal axle set in a frame. The tire is set spinning; because of the lubrication there is not much interaction with the frame, and so we can regard the tire as an isolated system, to a good approximation. The magnitude of its angular momentum depends on tire mass, rotation speed, and size. If you double the mass or double the speed, then the angular momentum doubles. If you double the tire radius, then the angular momentum quadruples.[10]

The laws of classical mechanics tell us that, for an isolated system, angular momentum is conserved. For many real systems, it is a good approximation to say that their angular momenta are conserved. For example, consider a tetherball that is attached to a rope that is wrapping itself around a vertical pole. As more and more of the rope wraps around the pole, the ball gets closer to the pole—it radius decreases. At the same time, the rotation speed increases; this is because of angular momentum conservation. Consider a comet approaching the sun.

It speeds up as it approaches, reaching maximum speed at its closest point of approach (perihelion), whipping round the sun, and then slowing down as it recedes into the distance. In this example, the conservation of angular momentum is enshrined as Kepler's second law of planetary motion (figure 8.6).

A third example of conservation of angular momentum is an often-cited case: an ice skater performing a spin. Initially, she is spinning on one leg and is shaped like the letter *T*, with her body and one leg

Figure 8.6 Angular momentum conservation: a universal principle, like energy conservation. (*a*) Kepler's second law of planetary motion, which is a statement of angular-momentum conservation for a body orbiting the sun, *s*. The area swept out by a line connecting the orbiting body and the sun is the same everywhere around the orbit (ellipse) for a given time interval. So if it takes a month to move from *a* to *b*, then it takes a month to move from *c* to *d*, as the areas (*gray*) are the same. Thus orbital speed increases as the body gets closer to the sun. (*b*) In 2007, a powerful F5 tornado struck Elie, Manitoba, Canada. Tornadoes also illustrate angular momentum conservation: wind speed is faster lower down the funnel, where the radius is smaller. ([*b*] Photo by Justin Hobson)

horizontal and the support leg vertical. Then she moves to reduce her rotation radius by moving her body to the upright position, lowering her leg, and holding her arms by her side or above her head. As a consequence of angular momentum conservation, her rotation speed increases significantly.

***

Now let's look at those dramatic vortex phenomena—many of which are classified under the apposite, if untechnical, label of *whirlwind*. We have seen how large-scale cyclones and anticyclones develop: low-/high-pressure differences at different points on Earth's surface lead to winds directly toward/away from the low-/high-pressure center; these winds get deflected to the right (in the Northern Hemisphere) by the Coriolis force, and so circulation patterns develop as in figure 8.3. For a tropical cyclone, the air spirals in toward the center, and so, because angular momentum is approximately conserved in this case, the circulation speed increases.

Why are vortices so common in fluid flow? For example, when water drains out of a sink, why does it usually spin, clockwise or counterclockwise, rather than simply drain without spinning? The bald answer is that any slight circulation that exists in the fluid is exaggerated as it flows to the drain, due to angular momentum conservation. But also a rotating fluid flows more efficiently down the drain when it is rotating. We can think of a rotating upflow of air, somewhat loosely, as an upside-down drain that is the most efficient way of moving air vertically through the atmosphere.[11]

There are many types of vortices in meteorology on scales that vary from centimeters to thousands of kilometers and from a few seconds to many days. These vortex phenomena have many different causes: the devil is in the details. To gain insight into vortex weather phenomena, we will begin with a devil and work our way up from there.

Dust Devils

These little beasts are the smallest-scale atmospheric vortices, with the shortest lives and the least energy. They are very common; almost certainly, there are hundreds of invisible dust devils for every one that is seen—if the ground they pass over contains no dust, then there is nothing to make them visible (though they can sometimes be made visible by forest-fire smoke or flame or by power-plant steam). To explain dust devils in terms of the simple physical laws and principles we have already discussed, I am going to have some fun by turning you into a neophyte weather god (temporarily, I wouldn't want you to get carried away). As such, you have the capability for rustling up certain atmospheric conditions, and you decide to play with this new power by following a simple recipe learned in weather-god school for making dust devils. Here's your recipe for a dust devil:

- One unit of wind shear
- Two units of unstable air near the surface
- Mix on a hot surface sprinkled with dust, such as a desert

Here's how it works. A hot desert causes very unstable air near the surface. This air rises and accelerates upward. As it rises, it pulls dust up with it, making the process visible. More important, the wind shear causes the rising column to rotate. Say that wind (light winds work best here) near the surface is from the south, whereas a few meters above the surface it is from the west. The rising column of air begins to change direction under the influence of this shear. It doesn't take much to get air to circulate, and soon a circulation of the whole column of rising air occurs: it spirals upward. As it accelerates, the spiral is tightened, meaning that the radius is reduced. So conservation of angular momentum tells us that circulation speed increases—the dusty air spins faster. You have made yourself a dust devil.

The radius of a dust devil can range from a few centimeters to 100 meters (say, from 6 inches to 300 feet), and the height they reach can be as little as 10 meters above the ground to 1,500 meters (say, 30 feet to 1 mile). Rotational wind speeds increase with size: small dust-devil wind speeds might be 40 kilometers (25 miles) per hour, whereas large ones can be twice as fast. Small devils may persist for a few seconds, while large ones can last an hour or more. A 10-meter-diameter dust devil pulls heat up from the surface at a rate (it has been calculated) of 50 kilowatts. Another way of estimating dust-devil power is by viewing it mechanically, instead of thermodynamically. Assuming a 3-millibar pressure reduction in the fully developed vortex, the total upward force inside the vortex is about 2.4 tons. From this force, we can (very roughly) estimate the mechanical lifting power of the dust devil—the power it expends in raising air by means of the updraft—by guessing a modest updraft speed of 4 meters per second (9 miles per hour), in which case we obtain a figure of 95 kilowatts. Thus, however we estimate it, the power of a small atmospheric vortex is tens of kilowatts.[12] A large dust devil is shown in figure 8.7.

Figure 8.7 A large dust devil in the Arizona desert. (From NASA, "Phantoms from the Sand: Tracking Dust Devils Across Earth and Mars," October 7, 2005, http://www.nasa.gov/vision/universe/solarsystem/2005_dust_devil.html)

Occasional damage is done by dust devils, with few fatalities. The strongest is as powerful and damaging as the lightest tornado. A significant fraction of atmospheric dust arises, literally, from the action of countless dust devils. Most dust devils put in an appearance in the early afternoon when the ground is hottest. Unlike tornadoes, dust devils arise under clear skies when the weather is fine.

TORNADOES

Tornadoes are the big brothers of dust devils—or perhaps more distant relatives, because their origins are very different. In fact, the origins of tornadoes are variable and not yet perfectly understood, though much theoretical progress has been made over the past 40 years. Most tornadoes are spawned from *supercells*, which we met in chapter 7. A tornado is, by definition, a column of rotating air between cumuliform clouds and the ground. It is born in the cloud and reaches down to the ground. It is usually visible as a funnel cloud because of the condensation of water vapor or the presence of debris—which can be anything from dust to car hoods—picked up from the surface.

Most supercells do not develop tornadoes, though most tornadoes—especially the very dangerous ones—spring from supercells (exceptions include *gustnadoes* and *waterspouts*). Tornadoes are as American as apple pie. Indeed, they are more American than apple pie; I remember plenty of apple pie as a kid in England, but I don't recall any tornadoes.[13] Large tornadoes *can* be found on all the continents of the world, except Antarctica, but are most common by far in Tornado Alley, a swath of midwestern states from the Gulf of Mexico north to almost the Canadian border, with special emphasis on Texas and Oklahoma. On average, two tornadoes per day touch down somewhere in the United States, though, like the tornadoes themselves, the annual figures fluctuate wildly. The strength of these tornadoes varies by many orders of magnitude. An important measure of strength is given by the Enhanced Fujita scale, which is based on the damage done by a tornado as assessed by the debris and destruction it leaves in its wake.[14] EF0 tornadoes are weak and do little damage; EF5 tornadoes are deadly and very destructive. These are the ones that can hurl cars, topple trucks, drive flying splinters into tree trunks, peel asphalt off highways, turn houses into matchsticks, and kill people.

Because tornadoes are not fully understood, and because they can be formed in different ways, meteorologists cannot yet reliably predict when they will arise. Certainly, they are generated mostly in supercell thunderstorms; there is a correlation between the severity of the storm and the chance that it will spawn tornadoes, but the correlation is weak. Thus some strong supercell storms generate no tornadoes or generate them only during part of the duration of the storm or only over part of the area. Other, weaker supercell storms can spawn very powerful tornadoes.[15] Only 2% of tornadoes are the very damaging EF4 to EF5 types. Usually, the wide (*wedge*) tornadoes are more powerful than the thin (*stovepipe*) tornadoes, which have a nearly constant radius from cloud base to ground, but again the correlation is weak. Often tornadoes that are spawned from hurricanes arise along an axis that is directed west-northwest to east-southeast of the center, in regions of strong vertical wind shear. Many tornadoes track from west or southwest to east or northeast (though they can wander unpredictably and even double back).

More than 90% of tornadoes rotate cyclonically (that is, counterclockwise in the Northern Hemisphere and clockwise in the Southern Hemisphere), as do the supercells from which they arise. The reason for this asymmetry is *not* the Coriolis force (if it were, there would be no anticyclonic tornadoes) because on the scale of tornadoes—a few miles at most—the Coriolis force is very weak indeed. Even the rotation of supercells themselves is not a direct consequence of Coriolis forces, because these rotating thunderstorms cover distances of a few tens of miles—again too small. (Consequently, anticyclonic supercells sometimes form, though they rarely give rise to tornadoes.) Only the much larger hurricanes, discussed in the next section, are reliably cyclonic due to the Coriolis force. It is worth mentioning that tornadoes are only a small part, an insignificant appendage, of the supercells that give rise to them. The power of a supercell is vastly greater than that of a tornado. (The insignificance and relative weakness of tornadoes may be disputed by a small sentient being on the ground whose house is about to be clobbered by an EF5.) Of course, we appreciate the power of tornadoes more than that of the storms above them because tornadoes reach down to our level, where we live.

Why, of all the places on Earth, is Tornado Alley located in the central United States? The one-line answer is: warm, moist air from the Gulf of Mexico meets cold air from Canada and dry air from the Rockies. This statement is a gross oversimplification; it points to a geographical preference and gives rise to the necessary meteorological conditions for tornadoes, but it is insufficient to explain the relatively rare phenomenon of tornadoes. Is there a recipe that a neophyte weather god can follow to whip up a tornado? There are likely several, but meteorologists do not have any of them. The recipes are probably complicated—too advanced for a neophyte, so you will have to stick with dust devils for now. We do know some of the main ingredients of one recipe, however:

- A supercell (a strongly convective thunderstorm that rotates)
- Vertical wind shear
- Warm ocean water

The one-liner provides these ingredients, given our simple physical laws and meteorological fundamentals. Thus the warm Gulf of Mexico provides the energy that powers supercells and the tornadoes that they spawn. This warm, moist air is blown north, where it encounters cooler air (perhaps from Canada) that forces it upward. Dry air from the Rockies provides the vertical wind shear. If conditions are right (and May seems to be the month when conditions are most likely to be right), then the moist air rotates (cyclonically) as it rises and a supercell forms.

Weak tornadoes may persist for only a few seconds, whereas strong ones last for an hour or more. They often peter out if they run into a region of precipitation. The strong ones take more stopping than the weak ones. We can obtain a rough estimate of the lifting power of a tornado in the same way as we did for dust devils. First note that in terms of pressure difference, the fast updrafts of supercells translate into a huge pressure difference of 100 millibars or more.[16] That is, the pressure inside a supercell-spawned tornado funnel is around 100 millibars less than the pressure outside. Such funnels can be 0.8 kilometer (0.5 mile) wide or more; the lifting force for a tornado with these parameters is about 5 million tons. Given updrafts of 160 kilometers (100 miles) per hour, that converts into a supercell-tornado power in the region of 200 gigawatts. (To put this figure into human perspective, the largest hydroelectric power station in the world—the Three Gorges Dam in China—produces 20 gigawatts.)

The literature on tornadoes contains considerable discussion about the cause of their (mostly cyclonic) rotation. Tornadoes usually occur at the interface of the updraft and downdraft regions—most often the rear flank downdraft, but sometimes it is the forward flank. This fact strongly suggests that vertical wind shear is giving rise to circulating tubes of air. These tubes are initially aligned horizontally but are then turned on end by the updraft. Warm air enters the low end and joins in the circulation, causing the tube to grow downward, perhaps to the ground. This method of generating tornadoes is sketched in figure 8.8.[17] It applies more generally than I have suggested: a horizontal wind shear will also give rise to tubes of circulating air, as shown in figure 8.8*a*. The clever aspect of this theory is that, combined with the common occurrence of veering winds,

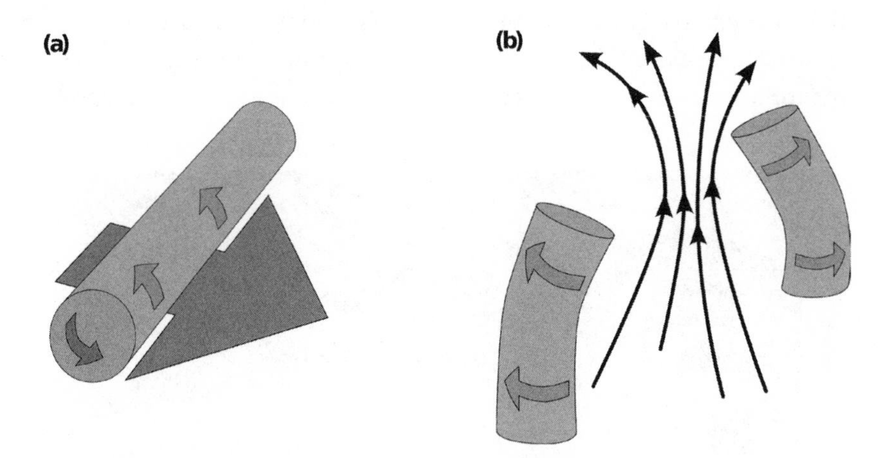

Figure 8.8 Cyclonic circulations. Here is one theory explaining the common occurrence of cyclonic circulations in small- and medium-scale weather systems. (*a*) Wind direction near the surface is to the right, but higher up it is to the left. In between, the air develops a circulation pattern, becoming a horizontal tube. It's like rolling a cigarette. Note that exactly the same circulation tube would arise if the wind on the right side was directed vertically up, and the wind on the left vertically down, as occurs at the interface between updraft and downdraft regions of a supercell. (*b*) The tube enters a region of updraft and is pulled out of shape into an upside-down *U*. It breaks, resulting in two more or less vertical circulations, one cyclonic and the other anticyclonic. In a veering wind (*large arrow*) the updraft is blown toward the cyclonic circulation, leading to a mesocyclone or to a tornado (that is, rotating updraft). Without the energy provided by the updraft, the anticyclonic circulation peters out.

it naturally leads to more cyclonic than anticyclonic circulations—be they dust devils, tornadoes, or mesocyclones. Also it is consistent with the observation of *multivortex* tornadoes—that is, two or more subvortices simultaneously orbiting the center of a larger circulation.

Supercells are rare, and tornadoes rarer, because of the unlikely conspiracy of wind shear and veer that is required to give rise to the necessary circulation.[18]

HURRICANES

When they occur over the Atlantic Ocean, or over the northeastern Pacific Ocean, they are called *hurricanes*. Over the northwestern Pacific, they are *typhoons*. Over the South Pacific and Indian Oceans, they are named *cyclones*. I am writing this book in North America, and so, to avoid confusion, I will call all of them hurricanes unless I mention them by name. In figure 8.9, we see a hurricane from above.

Figure 8.9 This is the best image of a hurricane that I can find that illustrates the huge scale of these storms. Typhoon Nabi lasted for 12 days, caused $972 million in damage over a large swath of East Asia in September 2005, dropped 132 centimeters (52 inches) of rain on Japan, produced 177 kilometer (110 mile) per hour winds, and generated a 75-millibar pressure reduction in its eye. This image of the category 4 storm captures its cyclonic rotation, eye wall, and size. (From NASA, International Space Station Imagery, September 3, 2005, http://spaceflight.nasa.gov/gallery/images/station/crew-11/html/iss011e12347.html)

Note that we have reached the opposite extreme of vortex power from the humble dust devils that we encountered earlier; dust devils are to hurricanes as firecrackers are to supernovas.

The basic building block of a hurricane is rainstorms—lots of them. Figure 8.10 shows you neophyte weather gods how to assemble these meteorological bricks to build the meteorological temple that is a hurricane. Knowing what you now know about the physical principles that apply in this field, you can see why hurricanes are structured this way. First, though, let's note the scale and power. These will provide insight into the wind structure of hurricanes and show us where the building-block rainstorms come from. The energy of a hurricane comes from the heat contained in the seawater over which they form. (In contrast to, say, a thunderstorm over land, where the energy comes directly from temperature differences among parcels of air.) If the water temperature exceeds 27°C (80°F) over a large stretch of ocean, then there is the possibility of a hurricane. Hurricane season in the Northern Hemisphere is June to November for this reason: these are the months of warm tropical waters. Hurricanes are huge—they are approximately circular with

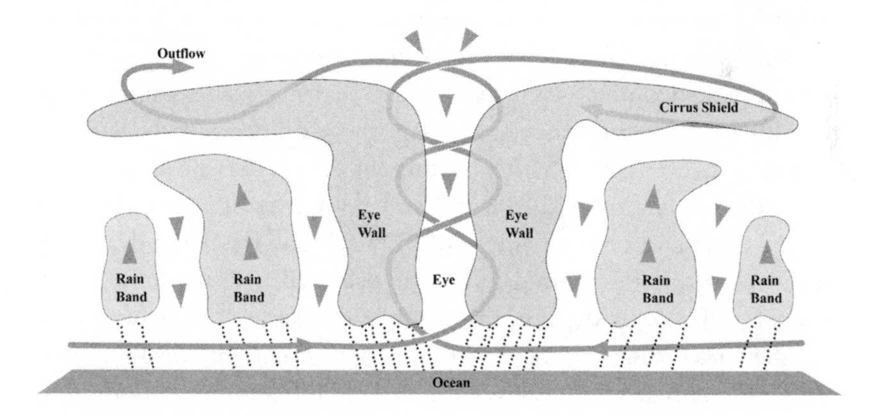

Figure 8.10 Hurricane structure. This cross section shows the eye wall and cirrus shield, which in three dimensions look like a mushroom cap with a hole (the eye) down the center. Bands of rain clouds circulate the eye wall. Warm moist air rises in these regions, and cool air falls between the rain bands. Air rises fastest at the eye wall and falls in the eye, which is clear and calm, though the ocean beneath the eye may be very rough. Most of the structure rotates counterclockwise, in the Northern Hemisphere (the outflow rotates anticyclonically, in this case clockwise).

an average radius of 665 kilometers (410 miles). The average rainfall inside this circle during the hurricane is 1.5 centimeters (0.6 inch) per day. From these observations, meteorologists can estimate the power of a hurricane. All that rain came from warm water evaporating from the ocean and being carried upward perhaps as high as the tropopause, where it condenses, releasing the latent energy of water vapor. This latent energy is known, and so from the total daily rainfall we obtain a power of 600,000 gigawatts. Thus an average hurricane consumes power at a rate that is about 30 times the world's entire electricity-generating capacity.[19]

Hurricanes extend from the ocean surface up to the tropopause, thus occupying the entire troposphere for more than 1.3 million square kilometers (500,000 square miles). Unlike that of tornadoes or thunderstorms, their width greatly exceeds their height, which means that hurricanes can be self-quenching: the storm-induced ocean cooling cuts off their energy supply (similar to the precipitation of some thunderstorms quenching the updraft). Environmental wind shear can also shorten the life of a hurricane, as can moving over land (which cuts off the energy source). The speed and direction of hurricane movement are well understood and can be predicted once the hurricane has formed. Obviously, the main purpose of prediction is to provide warning, given the damage that hurricanes can do once they hit land. Because they are so big, they can persist over land for several days even though their source of energy has been cut from under them. Hurricane Katrina claimed 1,300 lives in New Orleans in August 2005 and caused damage of $125 billion. The worst death toll in the United States from a hurricane occurred in Galveston, Texas, where 8,000 people died, mostly from storm surge.[20] Hurricanes are classified according to the Saffir–Simpson scale; the weakest (category 1) have sustained peak wind speeds of 120 kilometers (74 miles) per hour, whereas the strongest (category 5) exceed 185 kilometers (115 miles) per hour. Such winds push water: a category 5 hurricane can produce storm surges of 6 meters (20 feet). The worst-ever loss of life from this cause is an estimated 300,000 people in 1970 in what is now Bangladesh, a low-lying country that has flooding problems even without the surges caused by tropical cyclones. It has been estimated that the damage done by a hurricane that hits land increases as the cube of wind speed.[21]

The structure of a hurricane is shown in figures 8.9 and 8.10. From above, we see a cyclonic spiral with a hollow center—the *eye*. A vertical cross section shows bands of rainstorms. Moving in toward the eye, wind speed increases, reaching a maximum at the *eye wall* before dropping to nothing inside the eye. Temperature increases from the outside in. Air pressure drops as we move in to the center to as low as 920 millibars. The lowest pressure occurs in the eye, with small eyes (16 kilometers [10 miles] across) being associated with the strongest winds (category 5) and large eyes (48 kilometers [30 miles] across) with weaker winds. Vertically, the strongest updraft occurs at the eye wall. These updrafts are powerful because the temperature difference between the ocean water at the surface and the tropopause 12 kilometers (7.5 miles) above it can be 100°C (212°F). Bands of updrafts and downdrafts occur at different distances from the center, as suggested in figure 8.10. The massive amounts of clouds generated by the updrafts spread out at the top of the hurricane as a cirrus cap, or "shield."

From what we have learned about weather physics, we can understand much of this hurricane structure. The spiraling toward the center is due to pressure differences and the Coriolis effect. The increasing wind speeds toward the center are due to conservation of angular momentum. The eye arises because there is a centrifugal barrier where pressure difference pushing air inward is canceled by centrifugal force flinging the air outward.[22] The low pressure at the eye wall is a result of the strong updraft. The heavy rain obviously results from precipitation caused by moist air rising and cooling; the rain occurs in bands because of circulation near the surface.

Here is a brief summary of hurricane evolution. Warm ocean water evaporates, heating the air above it and causing it to rise. A tropical storm forms. As it grows, air from afar is sucked in (read: pushed in due to pressure differences); its path is bent by the Coriolis force, and so the circulation becomes cyclonic. More and more of the atmosphere above the storm is warmed by rising air—by latent heat released from cloud formation—causing the storm to grow to the tropopause, where it spreads due to stratospheric stability; this is the cirrus shield. The rising air cools and water vapor condenses, causing heavy rain. The energy driving the hurricane—ocean heat—is removed if the storm is driven over land by environmental winds, and so it slowly dissipates.[23]

Front and Center

Vortex phenomena are dramatic examples of complex weather that arises from simple physical laws and principles. More common examples, responsible for everyday weather and prominent on your television screen every time you turn to the Weather Channel, are weather fronts.

Physicists can describe the movement of an 11-kilogram (25-pound) cannonball through the air with considerable precision; the equations that describe its motion are simple, and the number of degrees of freedom—the number of ways the cannonball can move—is limited (to six: it can move along the *x*, *y*, or *z* axes, and it can twist about these axes). But consider the same mass of air—about 9 cubic meters (323 cubic feet), at sea level. To describe its movement requires many, many more degrees of freedom because the air is fluid and not rigid like the cannonball. Each part of the air can move in different ways; each part can twist in different eddies from other parts. The boundary is not fixed: the air can expand or contract or change shape. This is why fluid dynamics is so much more complicated than mechanics. We have discussed a *parcel* of air being an unspecified amount but with common properties such as uniform temperature throughout. The size is small and is roughly that of a dust devil or a town. Now, once again, I must introduce to you another necessarily imprecise concept (readers who crave precision, perhaps watchmakers or pharmacists, may be dismayed by this imprecision, but unfortunately it's an unavoidable consequence of dealing with fluids). An *air mass* is a large body of air, similar to a parcel, but much bigger—say hurricane size rather than dust-devil size. An air mass might blanket half a continent or an ocean. It is characterized by being roughly uniform in certain key parameters, principally density and temperature. Two air masses are distinct if their densities or temperatures (or even humidities) are different; thus when they are contiguous, we can distinguish the boundary between them. This last point brings us in a roundabout way back to weather fronts.

A weather front is defined as a transition zone between air masses of different density. Density depends on temperature (and humidity, to a much lesser extent), so these important variables also factor in. The significance of fronts to meteorology was first worked out by Vilhelm Bjerknes and his team in Norway around the time of World War I.

(The term "front" was given by analogy with the wartime front lines that divided Europe at that time.) Fronts are largely a phenomenon of mid-latitudes, where cold polar air masses meet warm, humid equatorial air masses. The meetings develop into cyclones with associated low-pressure centers and weather fronts, as described here. Weather phenomena such as thunderstorms or fog or shifts in wind direction tend to happen at fronts.

A quick survey of the five types of fronts will be instructive, in light of the simple physical laws and phenomena that we have highlighted in this chapter (figure 8.11). These fronts are responsible for a large part of the weather that you see every day above your head and on your television screen.

1. A *cold front* moves the fastest (it is not easily stopped by the less-dense air in front of it) and usually in a direction from northwest to southeast in the Northern Hemisphere (southwest to northeast in the Southern Hemisphere); it is the transition region where cold air replaces warm air, so that, if you are standing on the surface when the cold front passes, you feel the air temperature drop. Other meteorological occurrences that characterize a cold front passing are increasing air pressure, heavy rain, and poor visibility, followed by reduced humidity and shifting winds. The cold front forces the warmer air in front of it to rise and cool, and if this warm air is moist then precipitation results—a by now familiar story. In the Northern Hemisphere, cold fronts usually lie to the west and south of a low-pressure region.

2. A *warm front* ambles slowly, often from southwest to northeast because it starts to the east of a cyclone (a low-pressure center). A warm front passing over you results in increasing temperatures and falling pressure, a sudden wind shift, drizzle or perhaps fog, and increasing humidity. When a cyclone forms, the positions of cold and warm front are typically as shown in figure 8.12*a*, and evolve as in figure 8.12*b*. (The symbols used in meteorology for the different types of weather fronts are shown in figure 8.12.) You will, perhaps without noticing, have seen this particular pattern on weather maps hundreds of times. It is common in North America because the continental geography gives rise to the same air masses in the same places again and again. Thus a *continental polar* air mass always forms over northern Canada in winter (it is often called an "Arctic high"), and a *tropical maritime* air mass over

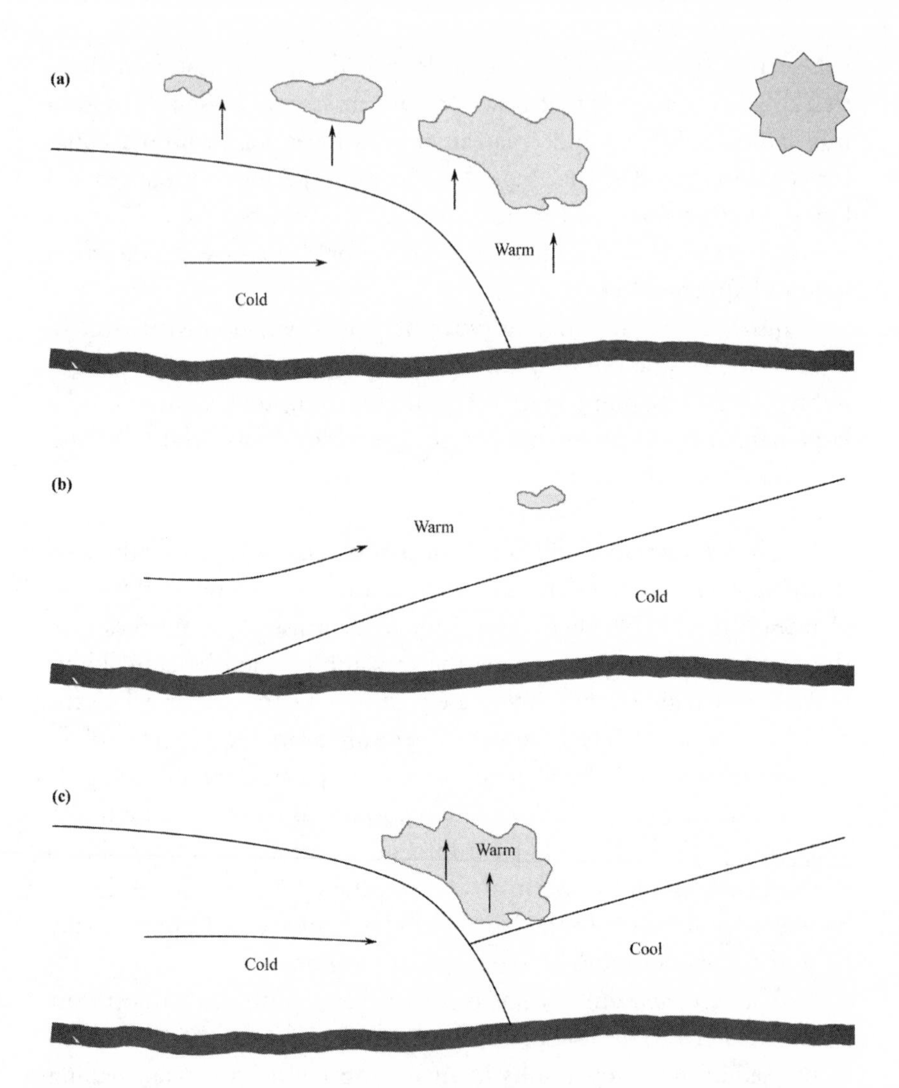

Figure 8.11 Weather fronts. (*a*) View from the side of a cold front penetrating warm surface air. The warm air is "wedged" aloft. If this causes water vapor to condense into clouds, then the sun might heat the warm air more than the cold air, thus enhancing the temperature difference between the two air masses. The steep camber of the cold front can cause the rapid rising of warm air. (*b*) Warm front: warm air rises over cold air. The camber is less in a warm front, and so the ascent is slower than for a cold front. (*c*) Occluded front: warm air is trapped between cold air and cool air. The cold front pinches off the warm air at the surface, forcing it aloft. The footprint of the warm air on the ground is sometimes called a *trowal.*

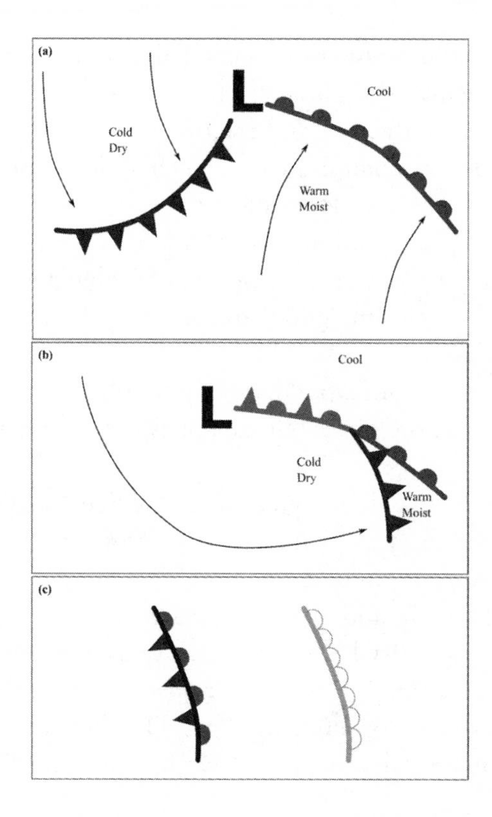

Figure 8.12 Putting up a front. (*a*) Plan view (usually superimposed on a map) of a low-pressure region (L), amid stationary cool air—say, in the Midwest. It pulls a cold, dry air mass down from Canada and a warm, moist air mass up from the Gulf of Mexico. Yes, these are the makings of a supercell. The transition region between moving warm air and slower cool air is the warm front; meteorologists and weather forecasters denote this front by a *line with semicircles on the side* toward which the warm front is moving. The cold front is denoted by a *line with triangles on the side.* (*b*) The cold front moves faster than the warm front and catches up to it, forming an occluded front (*semicircles and triangles*). (*c*) The symbols of other front types: stationary (*left*) and dry line (*right*).

the Gulf of Mexico. As we have seen, sometimes they meet up in Tornado Alley. In other parts of the world, similar geographic effects give rise to repeating air-mass formation; thus Europeans often hear their meteorologists talk of an "Icelandic low" or an "Azores high."

3. An *occluded front* occurs when warm air is trapped between two masses of cooler air. If the cool air regions approach each other, the

warm air is forced upward and is separated from the surface, resulting in a temperature inversion. It is easy to imagine cold air wedging itself underneath warm surface air and forcing it aloft. In fact, the physical process is much more complicated, so much so that meteorology has its own (long, inevitably) words to describe the process of front formation (*frontogenesis*) and dissolution (*frontolysis*). However, these complications are not important to us; the notion of wedging is a good enough paradigm and a much simpler one to convey.

There are two other front types that I will simply mention in passing—they matter to meteorologists but are not especially instructive.

4. A *stationary front* is, you guessed it, one that has stalled and does not move. A stationary front matters if a cyclone is on it: expect large amounts of precipitation for a long time.
5. A *dry line* is a boundary that separates moist air from dry air. It has its own symbol. Dry lines arise in the same areas repeatedly, again due to surface geography—for example, just east of the Rockies. Tornadic supercells can develop along dry lines. (The symbols used to denote stationary fronts and dry lines are shown in figure 8.12*c*.)

***

Differential heating of Earth's surface gives rise to horizontal and vertical air movement. This movement, acting under the influence of simple physical laws and complicated equations of motion, gives rise to much of our weather. Fluid is easily induced to circulate in vortices and eddies; circulating air plus wind shear give rise to complex and dramatic weather phenomena such as supercell thunderstorms and tornadoes. Weather fronts are common around mid-latitude cyclones, and are responsible for much of the (nonconvective) vertical air movement and so for much of our precipitation.

9

Weather Extremes

The New Normal

Black swans and tail events run the socioeconomic world.

Nassim Taleb

It is not just the socioeconomic world.[1] Of course, extremes by definition cannot be normal—the title of this chapter is hyperbole—but weather extremes leave such scars, on the physical world and within the people affected by it, that they figure prominently in any reckoning of the subject. In this penultimate chapter, we see why extremes of weather are becoming more common and look at the forms they have recently taken and will take in the future.

Feeling the Heat

There are two good reasons for starting off this chapter with a summary of the effects of heat waves. First, heat waves are the leading weather-related killer, at least in the United States. Second, the statistical nature of weather extremes is particularly evident when we examine heat waves. That heat waves are deadlier than hurricanes or floods is perhaps surprising, and is due to the wide area covered by a heat wave compared with the area affected by, say, a flood. A flood is dramatic and will attract media attention, whereas there is not much to show when you point a camera at a heat wave. Also, the fatalities due to a heat wave are often insidious and indirect.[2]

In 1995, Chicago sweltered under a heat wave. During the week of July 14–20, 485 people died directly from heat-related causes, and the total number of excess deaths (that is, over and above the number expected in a typical week) was 739. This figure is twice the death toll from the Chicago fire of 1871. Nationally, the decade 1992 to 2001 saw 2,190 deaths that were attributed to excessive heat, compared with 880 due to floods and 150 due to hurricanes during the same time period.

Internationally, these numbers are dwarfed by recent extreme heat waves in Europe, in 2003 and 2010. In figure 9.1, you can see that the heat wave of 2003 was centered on France, and, indeed, French people accounted for nearly half of the estimated 30,000 deaths that were attributed to this event.[3] Worse followed seven years later. The Russian summer of 2010 was torrid—the hottest in recorded history.

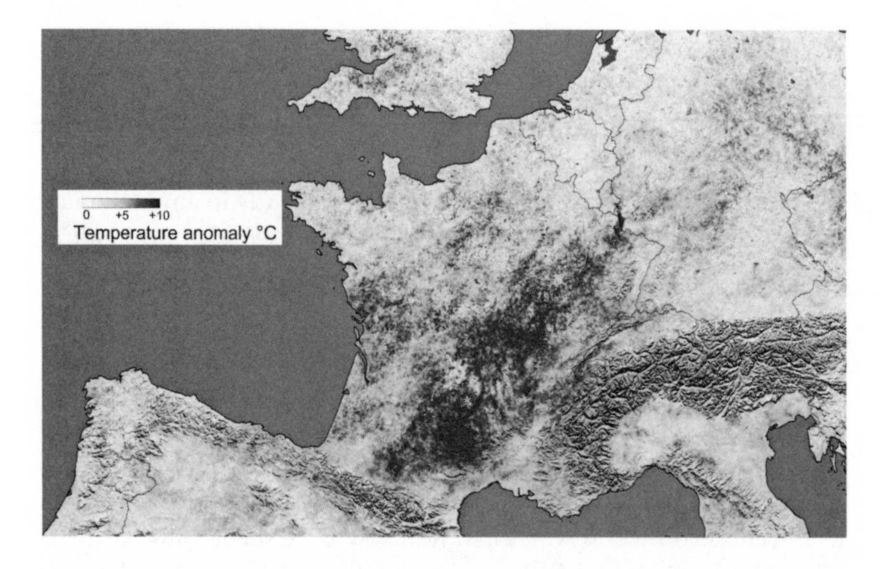

Figure 9.1 The heat wave in Europe in July 2003, which led to something like 30,000 extra deaths (when compared with the expected number), 14,000 of which occurred in France. (NASA, Earth Observatory, http://earthobservatory.nasa.gov/IOTD/view.php?id=3714; image courtesy of Reto Stockli and Robert Simmon, based on data provided by the MODIS land science team)

The thermometer topped 37.8°C (100°F) in Moscow on July 29. This may not seem like much to a resident of Tucson, Arizona, but for the people of a city renowned for its winter weather, it was brutal. As well as heat, the Russians had to deal with their worst drought in 40 years, with crop failures resulting from this drought and with high air pollution due to smoke from the large number of forest fires that arose. In total, some 55,000 people died.

In general, heat waves cause deaths by hyperthermia, maladies such as dehydration, heat cramps and heat stroke, damage to structures through heat expansion, increased frequency and severity of wildfires, drought, increased soil erosion,[4] and pollution (due to plant leaves reducing absorption through their leaves to cut down on water loss).

How are heat waves measured? The criteria vary from country to country. These definitions are often ambiguous and inconsistent. One definition is this: a heat wave is declared when the thermometer on three consecutive days exceeds the 90th percentile of maximum temperatures. The World Meteorological Organization declares a heat wave when the daily maximum temperature for five consecutive days exceeds the average maximum temperature by 5°C (where the average is defined as that of the years 1961 to 1990). Here we begin to see where statistics comes in and why they muddy the waters. These definitions have in common a critical temperature (though defined differently). In both cases, the thermometer just has to reach this temperature to record a day of extreme heat, but surely if the critical temperature was exceeded for five hours, then that should count more than if it was reached for only five minutes. Another factor in determining the severity of a heat wave is the minimum temperature at night. High minimum nighttime temperatures kills people as much as high daytime temperatures and also causes more fires. And what about humidity? The dry heat of the American Southwest is easier to cope with physiologically than the moist heat found in India.

However it is defined, it seems that both the frequency of heat waves and the areas covered by heat waves are increasing. During the three decades 1951 to 1980, heat waves covered between 0.1% and 0.2% of Earth's surface at any one time. During the next three decades, 1981 to 2010, the figure increased to 10%. It is expected that the percentage will increase again over the three decades from 2011.[5] The annual number of hot days has more than doubled since 1950. Of course, the manner in which heat waves are dealt with varies greatly from region to region, depending on national resources and infrastructure, as well as on what is locally considered to be hot—a Canadian would sweat in heat where a Mexican might not.[6]

Drought

The ancient state of Yemen (known as Arabia Felix [Happy Arabia] to the Romans) is likely to become a failed state in the next few years. This debacle will not be due to long-running sectarian tensions or to overpopulation, but simply because the country will soon run out of water. The problem of water shortage and the allocation of this dwindling resource spreads beyond Yemen into the greater Middle East, a region already unstable because of many ethnic and religious tensions between neighboring states. This part of the globe contains some 1% of the world's freshwater but 5% of its population. There have been disputes among Turkey, Syria, and Iraq over the Tigris and Euphrates Rivers; among Israel, Lebanon, and Jordan over the Jordan River; and among Egypt, Ethiopia, and Sudan over the Nile River.

Further afield, the always-shallow Aral Sea has decreased in size to 10% of what it was in the 1960s. This reduction is not due to natural water loss or even to climate change but is a direct man-made disaster: 50 years ago, the Soviet government decided to divert the two main rivers that feed the Aral Sea, the Syr Darya and the Amu Darya, for irrigation purposes. Reduced volume and increased salinity killed off most of the fish and all of the fishing industry, causing local economic hardship. Successor states that are suffering as a result of this water-management disaster include Kazakhstan, Uzbekistan, Turkmenistan, Tajikistan, and Kyrgyzstan. These nations have committed resources in an attempt to increase the volume of the Aral Sea; for example, Kazakhstan has built a dam that has led to an increase in water levels, a reduction in salinity, and a partial recovery of fish stocks. There is a message here: human environmental mismanagement can be reversed with clear thinking and engineering backed up by resources—and, most important of all, international cooperation.

Such cooperation needs to be timely and is perhaps in increasingly short supply as environmental pressures make freshwater scarcer or less accessible. The concept of *peak water* was defined in 2010. The notion arises from peak oil, which is the name given to the point in time at which the maximum extraction rate is reached—clearly such a peak will occur for a nonrenewable resource.[7] Introducing peak water suggests that water, too, is effectively a nonrenewable resource. Proponents of this view point to the fact that freshwater in lakes and in underground aquifers can be depleted (for example, the once-giant Ogallala Aquifer beneath the American Great Plains). There are perhaps three peaks: peak renewable water, which is the point of time when the entire renewable water flow across the world is being consumed by humans; peak nonrenewable water, when groundwater aquifers are overpumped; and peak ecological water, the point beyond which the cost of ecological disruption exceeds the value of the water to humans.

The notion of peak water emphasizes that, around the world, freshwater is becoming less available. It is anticipated that by 2025, 1.8 billion people will be living with insufficient water resources. The United Nations has warned that climate change harbors the potential for serious conflicts over water: "Water wars are coming."[8] Perhaps not just water wars: demographic pressures due to global warming are mounting. The changes we have experienced will continue and will vary from place to place over the globe. Generally, the dry regions north and south of the tropics will get drier, causing water and arable land shortages, leading inevitably (given the rising population of undeveloped and developing nations) to population moves away from these regions to the more affluent temperate latitudes. Political and perhaps military tensions will rise near the northern horse latitude[9] (that is, at the boundary between Hadley and Ferrel cells, you may recall, where deserts occur). We are already seeing increasing pressures on the southern U.S. border with Mexico as the number of illegal migrants from Central America grows each year, and in the southern European countries as migrant boats from Africa carry more and more people toward (they hope) better lives—if they survive their journey.

It is easy to make apocalyptic predictions about these demographic trends arising from environmental changes, and it is appropriate to mention them in a chapter about extremes. It seems to me a little premature, however, to be manning the barricades just yet in defense of water resources or farmland. Recall the progress made in turning around, partially at least and perhaps only temporarily, the desiccation of the Aral Sea. Note also that progress has been made in the Middle East water supply with an increased capacity there for desalination (and a consequential reduction in desalination costs per cubic meter of fresh water produced). In China, the last decade has seen massive improvements to water supply and sanitation for millions of people, helping to bring them out of poverty. All these changes have been technological—if (and only if?) these technological innovations can be matched by concomitant political cooperation between nations, then the nastier scenarios that arise from climate change extremes might be avoided.

A test case for the developed world of these ideas—one that is much closer to home for many readers—is a state that is much richer than Yemen and more politically stable, but that also has a dire water shortage. By the spring of 2015, California had been suffering for four years from its worst drought in more than a century. It affected more than half the state, and in April 2015 Governor Jerry Brown introduced the first mandatory water restrictions (and declared a drought state of emergency). Unfortunately, these restrictions did not apply to farmers, who consume most of the state's water supply as irrigation for the huge fruit, vegetable, and nut industry that supplies half the nation's needs. In March 2015, NASA predicted that California's reservoirs would dry up within about a year—no surface water is already a fact in the productive Central Valley. To compensate, farmers are digging unregulated wells deeper and deeper as the water table drops (it has fallen 15 meters [50 feet], to date). This loss of groundwater is unsustainable: it is already causing the surface to sink in many areas, by up to 0.3 meter (1 foot). Soon no groundwater will become the new normal in the Central Valley. It will be interesting to see how or if California can mitigate its water shortage without a long-term decline in agricultural productivity or quality of life. The state lifted its water restrictions in May 2016.[10]

In from the Cold

Having examined extremes of heat and of water scarcity and their possible consequences, in this section and the next we turn to exactly the opposite: extremes of excessively cold weather and too much water.

Very cold weather kills fewer people than other extreme weather phenomena, at least in developed countries where shelter and support are available.[11] It can be very costly in economic and financial terms, however, and extremely disruptive. The main risks to human health arise from frostbite and hypothermia. Economic and financial costs pile up during and after an extensive cold snap due to crop losses, extra feed for livestock and more livestock deaths, increased fuel consumption, frozen pipes, and extra house fires caused by additional heating. Furthermore, fire damage is often worse during a cold period because of the particular difficulties of fighting fires in cold conditions (burst water pipes, difficulty of access in heavy snowfall, and so on).

The severity of cold-weather extremes is generally distributed unevenly across a nation and differs from the distribution of heat-related extremes. Thus heat waves cause problems in urban areas where heat is trapped in and between buildings (the *urban heat island effect*), but the same effect reduces the severity of cold waves, which are worse in rural areas. The effects also vary with climate, so that a cold snap in Florida may destroy a fruit crop, whereas a more severe or more extended snap in Minnesota may do little damage.

Water, Water Everywhere

One-third of the annual number of meteorological disasters around the world arrive in liquid form—flooding. Floods kill a large number of people in the Third World and cause huge monetary damage in the developed world: one-half of those affected by meteorological disasters are the victims of flooding. Thus the worst natural disaster in world history, in terms of loss of life, occurred between July and November 1931 when the Yellow, Yangtze, and Huai Rivers in China flooded catastrophically: millions of people died—several sober estimates say as many as 3.7 million. More recently, 30,000 fatalities arose from severe flooding in Venezuela in December 1999. The 1,500 deaths in New Orleans that resulted from Hurricane Katrina, in 2005, were at least in part attributable to the levees breaking, which caused local water levels to rise 6 meters (20 feet). Flooding kills directly by drowning and indirectly by compromising sanitation and spreading water-borne diseases.

In monetary terms, flooding costs governments and insurance companies billions each year. In 2012, the U.S. state with the highest flood bill was New Jersey at $3.5 billion, with New York close behind. The

World Bank is sufficiently concerned about these rising costs that it now advises governments about mitigation strategies. Given these costs, it is unsurprising that those countries that can afford it choose to invest huge resources in flood prevention or mitigation, buildings dikes (levees), sluices, locks, dams, floodgates, and barrages, which defend against rising river levels due to heavy rainfall or snow melt or to rising sea levels. There are 3,500 miles of levees (Americans prefer the French word, as opposed to the Dutch) along the Mississippi; other river defenses in the developed world can be seen along the Po, Rhine, Loire, Vistula, Scheldt, and Danube Rivers, among others. Sea defenses include the Thames Barrier in London and coastal-flood-prevention works in the Bay of Fundy (which has the highest tides in the world), Vancouver, and of course the Netherlands. Fully two-thirds of the Dutch countryside is vulnerable to encroachment from the sea, which is held at bay by massive constructions such as the Zuiderzee Works (which turned the Zuiderzee, a saltwater inlet, into a freshwater lake and added 1,650 square kilometers [637 square miles] of land to the country) and the Delta Works (which the American Society of Civil Engineers has named as one of the seven wonders of the modern world [figure 9.2]). Climate change and urban developments pose challenges for such flood development-engineering projects, which are very hi-tech in many cases and involve detailed mathematical modeling of flood generation and prediction.

Low-lying coastal countries that are impoverished (for example, Bangladesh) or small (for example, Seychelles) fear rising sea levels and cannot afford the huge costs of defensive structures. Impoverished or developing countries with shantytowns on riverbanks suffer fatalities from rapidly rising water levels (or from mudslides, if the shantytowns are on hillsides) following extended periods of exceptionally heavy rains. Both these causes of flooding are expected to become worse and more frequent due to ongoing climate changes—the number of people affected by river flooding worldwide is expected to triple by 2030.[12]

Storms

Let us look at extreme storms. Here I will restrict attention mostly to hurricanes—that is, storms that arise in the North Atlantic—which are well documented and occur close to home for many readers.

Figure 9.2 One of three movable barrier sections of the Oosterscheldekering (Eastern Scheldt storm surge barrier), just one part of the extensive Delta Works in the Netherlands. These works were constructed in response to devastating floods in 1953—the same year as flooding in England that eventually resulted in the Thames Barrier. (Photo by Vladimir Šiman, 2008)

Increasingly good records have been kept since the mid-nineteenth century. In figure 9.3*a*, we see the deadliest hurricanes to make landfall in the United States. The first thing to note about hurricane fatalities is that they reflect two aspects of the storm: its severity and its location and year. In particular, long ago we were not well prepared to deal with these natural disasters, which sadly is still true in many places around the world. Consequently, severe hurricanes from decades ago, or from underdeveloped regions today, tend to cause much higher death rates than storms of the same severity that hit a developed country today. Thus we note from the figure that four of the top five hurricanes occurred early, before the naming convention for hurricanes began,[13] and one (Mitch) was centered in Honduras in Central America. Katrina stands out as the exception. Over and over as we go through

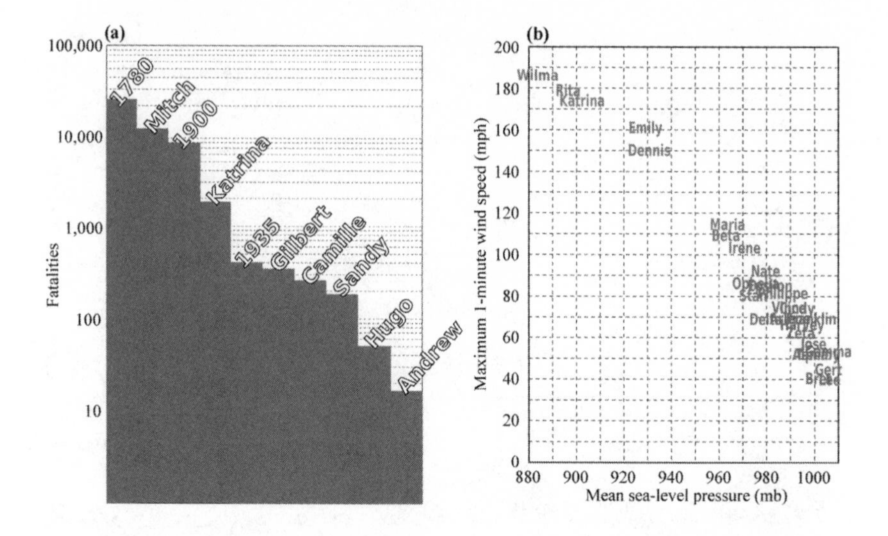

Figure 9.3 Hurricane data sliced two ways. (*a*) The most severe hurricanes of all time, as measured by the number of people killed. Mitch occurred in 1998; Katrina, in 2005; Gilbert, in 1988; Camille, in 1969; Sandy, in 2012; Hugo, in 1989; and Andrew, in 1992. (*b*) Hurricanes that arose in 2005. The plot is of maximum wind speed (1-minute average) versus mean sea-level pressure at the storm center.

the statistics for these storms, we find that the main cause of death is due to storm surge—not flying debris or collapsing trees or buildings, but seawater that is pushed inland by the exceptionally strong winds. Such was the case with Katrina, as we are about to see.

One very good reason why a strong hurricane is not necessarily deadly is that hurricanes arise over warm seas and fully two-thirds of them do not make landfall.[14] Perhaps a more objective estimate of the intensity of a hurricane can be obtained by looking at the minimum atmospheric pressure at the center and at the maximum recorded wind speed (these two variables are not independent, as you can see from figure 9.3*b*, and as you might anticipate from the discussions of circulating storm physics in chapter 8). In figure 9.3*b*, we see that high wind speeds go with low central pressures. Note that, again, Katrina stands out.

Hurricane Katrina was not the strongest on record, but it was very large and it made landfall not once, but twice. Tropical storm Katrina arose on August 23, 2005, near the Bahamas, and headed west, picking up strength enough to be reclassified as a hurricane just before making landfall in southern Florida, where it did a great deal of damage and killed 14 people. Katrina continued west into the Gulf of Mexico before turning north, making landfall again near New Orleans—a total of 1,577 lives were lost in Louisiana. Weakening as it headed farther inland, Katrina turned northwest before petering out in Ohio on August 31. The storm surge on the coasts of Louisiana and Mississippi reached 8.5 meters (28 feet)—the highest ever recorded on the shores of the United States—and penetrated an average of 10 kilometers (6 miles) inland (twice that distance near bays and rivers). Levees broke almost everywhere, and 80% of the city of New Orleans was flooded. Much blame and finger-pointing resulted from the failure of storm-protection structures—the levees were later rebuilt to a higher standard.

Katrina was a very widespread storm. More than 233,000 square kilometers (90,000 square miles) of land were impacted. The storm reached out, in less severe form, as far afield as the Bahamas, Cuba, and eastern North America (the St. Lawrence River Valley in eastern Canada received 10 centimeters [4 inches] of rain from Katrina). The final death toll may never be known: in addition to the 1,833 confirmed fatalities, there are 705 persons classified as missing. The monetary cost was the highest of any hurricane by far: $108 billion.[15]

Winds of (Mis)fortune

Strong winds—and powerful gusts, especially—apply great force to man-made structures and can tear them apart. A steady wind can push over a building or a high-sided vehicle; gusts shake a structure and test its strength in several different directions. Very high-speed winds can toss debris hundreds of meters, rip paving off road surfaces, and tear the bark off trees. Extreme wind speeds—the record is 512 kilometers (318 miles) per hour from a tornado that hit the suburbs of Oklahoma City on May 3, 1999—can throw vehicles onto the roofs of buildings and reduce cities to matchsticks. Moore, Oklahoma, looked like one of the bombed-out German cities in 1945.[16]

European windstorms are extratropical cyclones that start off life as nor'easters around the New England coast. They hammer northern England and Scandinavia quite commonly in the winter, but can veer south to sweep across any western European country. They are the cause of the highest insurance losses of any type of natural disaster after U.S. hurricanes.[17] High winds lead to flooding as seawater is pushed inland, to loss of electrical power as pylons are blown over or power stations are shut down for safety reasons, to interruption in marine traffic, to road and rail closures, and to trees being blown onto roads and roofs. Thus the Great Storm of October 1987 blasted England and France. It brought down an estimated 15 million trees (including one-quarter of those in Kent and in Brittany) and led to the loss of electrical power for several hundred thousand people for days. The wind speeds in this storm rivaled those of Atlantic hurricanes, though the nature of the storm was very different. Insurance losses in the United Kingdom amounted to £2 billion, and the damage in France was estimated at 23 billion francs.[18]

On average, four or five windstorms strike Europe each winter. Three storms in December 1999 (named Anatol, Lothar, and Martin) inflicted $13.5 billion (indexed to 2012) damage and led to 150 deaths.

Climate Attribution

In the news media, there is a flurry—if not a storm—of articles reporting extreme weather events, in many cases attributing these events to climate change. Of course, survivors of the statistical ruminations of chapter 6 will know that we cannot easily attribute any particular weather event to a change in climate, just as we cannot attribute any single data-point extreme value to a change in average value. If your local weather station predicts that 5 millimeters (0.2 inch) of rain will fall in your neighborhood tomorrow morning and you get 6 millimeters in a bucket in your garden, what does that tell you? Is the excess due to random fluctuation, or does it reflect an increased average? If a later weather report says, "We forecast 5 millimeters of rain, and that's what you got," then you know that your bucket level was a local fluctuation. If a later report says, "We forecast 5 millimeters, but you got 7 millimeters," then, again, you know that you got a fluctuation. Fluctuations generally drown out changes in the average of a statistical ensemble, so it is next to impossible to attribute an unexpected individual result to a change in the average.

But what about a series of extreme weather events? For example, the average number of Atlantic hurricanes that arose over the years 1851 to 2010 was 5.4 per year, but the annual average over recent years, from 1995 to 2010, was 7.9 per year. More generally, there has been a well-documented increase in the number and severity of extreme weather events over recent decades. Consider 2014: it was a record warm year and produced heavy floods in Kashmir; droughts in California and Brazil; a tropical cyclone, Hudhud, that led to a deadly blizzard in east-central India; super-typhoons Neoguri and Hagupit (category 5) in the Far East; twin EF4 tornadoes in Nebraska (June 16); and 2 meters (78 inches) of lake-effect snow in Buffalo. *National Geographic* listed the most extreme weather events of 2013[19]—this practice is now common in popular newspapers and magazines and helps to raise public awareness of meteorological trends, but always in the context of global warming and climate change. Can a series of events reflect underlying trends?

Yes they can, but the problem here is the relatively short length of the series—perhaps three decades. Nevertheless, the evidence is mounting, suggesting that climate change is responsible for the increase in number and severity of extreme events. Insured losses from natural disasters in the United States amounted to $36 billion in 2012, up 50% from the mean value of the previous decade. The average rainstorm has produced more rain these past few years than it did 40 years ago (perhaps because, due to the increase in global atmospheric temperature, the atmospheric water vapor content has increased): "What's going on? Are these extreme events signals of a dangerous, human-made shift in Earth's climate? Or are we just going through a natural stretch of bad luck?"[20] More than likely, both are true. Climate models tell us that the number of extreme events will increase, though—let me say it again—we cannot easily ascribe any single event to climate change. An analogy exists with epidemiology: a spike in leukemia cases near nuclear power plants, or a higher than average incidence of lung cancer among smokers or of pneumoconiosis among coal miners—these trends indicate statistical correlations, but much work needs to be done to establish cause and effect. The links between smoking and lung cancer and between pneumoconiosis and coal mining are now well established, and the link between extreme weather events and climate change is on its way toward being established.[21]

It *may* be possible to attribute some individual events to climate change, if the event has a signature. Think of a deer carcass in the forest. You may know that in your area, 30% of such carcasses are eaten by wolves and 70% by bears, which, without further data, means that all you can do is estimate the probability that a given carcass was eaten by a bear as 70%. But you can do better if you see paw prints around the carcass. The weather equivalent of such sleuthing is called *climate attribution*; it is the forensic investigation of individual weather events for telltale signs of climate change. The technique is controversial; its proponents claim to have found such a link for some events such as the European heat wave of 2003, but not for others such as the Russian heat wave of 2010. Twenty research groups examined the causes of 16 extreme weather events from 2013 and arrived at the following conclusions:[22]

1. Human-caused climate change greatly increases the risk for extreme heat waves.
2. Natural variability plays a much larger role in other events such as droughts and floods.
3. For three of the severe 2013 storms studied, there is no anthropogenic signal.

As always in science research, more data is needed to pin down what is going on with increasing levels of confidence.

***

The number and severity of extreme weather events are increasing around the world. In developed countries, heat waves are the deadliest weather phenomenon; they are increasing in frequency, likely due to anthropogenic climate change. Floods kill more people in the developing world. These and other weather extremes (drought, cold, hurricanes, and windstorms) are exacerbated by blocking patterns (box 9.1), which stall the normal evolution of weather systems. The attribution of extreme weather events to climate change is a developing field of study, currently controversial.

Box 9.1
Blocking Pattern

At temperate latitudes, weather normally moves from west to east, pushed along by the prevailing winds (see figure 3.6*b*). Sometimes this progression gets stalled, or blocked, by an atmospheric pressure disturbance known as a blocking pattern or blocking high. The mechanisms are not fully understood, but meteorologists know to look for the development of blocking patterns in the upper atmosphere, when the polar jet stream shifts from the normal zonal (latitudinal) flow to strong and persistent meridional (longitudinal) flow (see figure 3.8*c* and *d*). These shifts in jet-stream behavior occur when the North Atlantic is warmer than normal, and a blocking event occurs when the jet stream pinches off a large mass of air that normally flows. These kinks in the usual flow pattern last from five days to several weeks, and they stall the movement of weather systems. The result is an unusually long spell of warm weather or cold weather or dry weather or rain—whatever the weather happens to be when the blocking pattern establishes itself. In the Northern Hemisphere, blocking patterns occur in the spring over the eastern Pacific and the Atlantic and in the winter over western Russia and Scandinavia. They usually are the result of high-pressure ridges because these are most widespread and move more slowly than low-pressure air masses.

There are several types of blocking pattern. The *omega block* and *rex block* are sketched in box figure 9.1*a* and *b*. There are also *ring of fire*, *split flow*, and *cut-off low* blocks.

In my region of the Pacific Northwest, we experienced an extremely mild winter of 2014/2015, with weeks of sunshine and low winds. More serious was the rex block that established itself over Russia in the summer of 2010. The high-pressure region of the block (see box figure 9.1*b*) caused a heat wave, as we have seen. To the south, the low-pressure region of the same block led to an unusually strong monsoon season in Pakistan.

The stagnation in weather due to blocks naturally leads to extreme weather: dry periods extend into droughts, wet periods generate flooding, and so on. Weather forecasters are obviously keen to know when blocks are developing because they can lead to serious consequences and because they tell us that the weather tomorrow is going to be the same as it is today, for several tomorrows. Extreme weather and blocking patterns have been increasing in frequency over the past decade (more than doubling during summer months). Climate models are not very good at predicting the frequency of these patterns, often underestimating the number that arise. For what it is worth, several such models

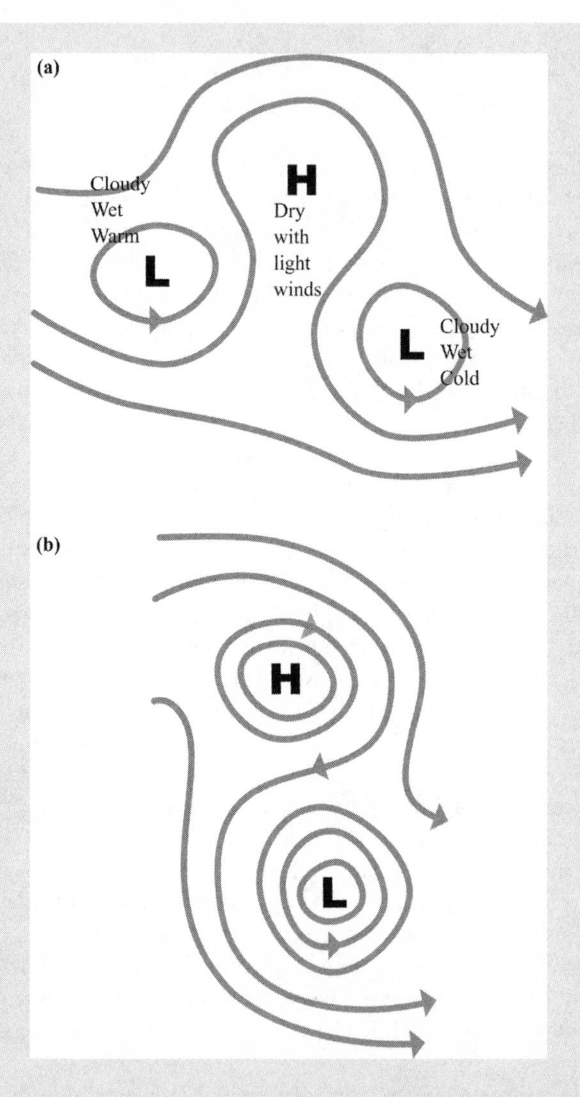

Box Figure 9.1 Blocking patterns. (*a*) Omega block: a ridge of high pressure blocks the normal movement of weather from west (*left*) to east. The lines represent isobars (along which wind flows due to the Coriolis effect, you may recall) at an altitude at which atmospheric pressure is usually half that at the surface. (*b*) Rex block: the sinusoidal flow of air between high- and low-pressure regions slows its progress eastward. (*c*) "Snowmageddon" on the East Coast of the United States due to a blocking pattern, February 11, 2010. A block in March 2013 in this region brought unusual heat; a year later, another brought cold to Washington, D.C.—a "freaking cold vernal equinox" (R. Grow, "Record Blocking Patterns Fueling Extreme Weather: Detailed Look at Why It's So Cold," *Washington Post*, March 21, 2013). ([*c*] NASA, Earth Observatory, http://earthobservatory.nasa.gov/NaturalHazards/view.php?id=42680&src=nha; image by Jeff Schmaltz, from NASA's *Terra* satellite)

(c)

predict a reduction in blocking-pattern frequency for the last quarter of the twentieth-first century. Hopefully before then, the models will have improved sufficiently for us to get a better handle on this extreme-weather precursor.*

*Extreme weather and blocking patterns are discussed in popular accounts by D. Carrington, "Extreme Weather Becoming More Common, Study Says," *Guardian*, August 11, 2014; and R. Grow, "Record Blocking Patterns Fueling Extreme Weather: Detailed Look at Why It's So Cold," *Washington Post*, March 21, 2013; and more technically in D. Coumou et al., "Quasi-resonant Circulation Regimes and Hemispheric Synchronization of Extreme Weather in Boreal Summer," *Proceedings of the National Academy of Sciences of the USA* 111 (2014): 12331–12336; S. Häkkinen, P. B. Rhines, and D. S. Worthen, "Atmospheric Blocking and Atlantic Multidecadal Ocean Variability," *Science* 334 (2011): 655–659; and T. N. Palmer, "Climate Extremes and the Role of Dynamics," *Proceedings of the National Academy of Sciences of the USA* 110 (2013): 5281–5282. For a popular account of the rising risk of extreme weather, see R. Harrabin, "Risk from Extreme Weather Rises," BBC News, November 26, 2013.

10

The World of Weather Forecasting

But who wants to be foretold the weather? It is bad enough when it comes, without our having the misery of knowing about it beforehand.

Jerome K. Jerome

Most of us pay attention to a weather forecast at the start of our day—simply to know what to wear or what to expect. For some people, however, knowledge of the upcoming weather is much more important. Truckers, bus drivers, ferrymen, and others who work in transportation need to know about driving conditions—snow or ice on the roads, fog affecting visibility, rain causing flooding, diversions or road closures due to weather conditions, and so on. Farmers like to know short- and long-range forecasts to help them decide whether to water a dry field of wheat or when to harvest corn, cover fruit trees threatened with frost, move stock to different pastures, or spray for bugs. Air-traffic controllers need weather information to decide when they must close an airport, when they can reopen one, or whether and where to divert a flight. Fishermen want to avoid the perfect storm. Foresters worry about wildfires and how to allocate resources during forest fire season. Retailers like to know the forecasts so they can restock to meet anticipated demands—for electricity generators, for example. Construction companies have to know about wind conditions and rainfall levels. Utility companies can operate more efficiently if they are able to anticipate demand—water during a drought or electricity during a cold snap or heat wave. Military operations are influenced by weather, and indeed many military organizations are active in weather data gathering and weather forecasting.[1] In addition, the meteorological services provide information to the general public about atmospheric conditions that lead to increased pollution (smog intensifying in valleys, the likely direction of smoke due to forest fires or of volcanic plumes).

Weather and climate affect almost everything that we humans do—one-third of the world's total economic output is influenced, in one way or another, by weather. The clothes we wear and the beer we drink depend on the state of the atmosphere. Even the best time to advertise soft drinks is deemed by marketing people to be weather dependent. Retail sales in general depend on temperature—prices of orange juice are influenced by hurricanes in Florida.[2] Bakeries decide whether to buy wheat futures based on long-range weather forecasts. The siting as well as the operation of energy infrastructure depends on weather—think of wind farms and solar-power plants.

Prediction: Forecasts Improving Rapidly

The need for accurate weather forecasts is increasing because the population of the world is increasing. The accuracy of these forecasts is improving as a result of

- Increased knowledge of the basic physics
- Increased data gathering
- Vastly increased computing capabilities

From time immemorial, farmers and mariners have cast anxious eyes at the skies, attempting to divine the mercurial ways of nature. By the seventeenth century, scientists had developed barometers and thermometers to assist with this enterprise. In the following century, systematic observations of weather began; they have spread around the world and increased in frequency and scope ever since. In these early days, weather prediction was based on little more than barometric pressure and current sky conditions. A major advance in weather forecasting came with the development of the electric telegraph in the mid-nineteenth century; for the first time, weather observations could be gathered from distant parts and assembled before the observed weather arrived. Theoretical work from the 1890s led to improved understanding of weather systems and, consequently, improved forecasting. Lewis Fry Richardson in England founded the science of numerical weather prediction (NWP) with a key publication in 1922. Inadequate computing resources (Richardson's calculations were performed by hand) and input data meant that early forecasts were often late or wrong—the many and complex calculations could not be done in real time until after World War II. Even then, an experienced forecaster performed better than calculated weather predictions until the 1950s. A big step forward in data gathering was made with the deployment of remote sensors following World War II; weather radars and then infrared cameras were developed. More recently, weather satellites have expanded our perspective and permitted meteorological information to be gathered on a planetary scale, providing snapshots of synoptic (large-scale) weather systems. From the 1990s, automated weather stations on land and sea have provided more and better data, as we have seen. The scale of meteorological forecasting can be seen from the size of the National Weather Service (NWS), the sole government agency for civilian weather forecasting in the United States—it has 122 forecasting offices.

Over decades, prediction capabilities improved, though even today the human input to meteorological forecasting is still significant. There are many measures of these improvements. Thus a three- to four-day forecast made in 2002 was as accurate as a two-day forecast made in the 1980s. According to the Met Office, the national weather service in the United Kingdom, its four-day forecasts are now as accurate as its one-day forecasts were 30 years ago.[3] Another source claims that seven-day forecasts are now as good as three-day forecasts were 20 years ago. However we cut and slice it, prediction accuracy is improving. No longer can we nod at Patrick Young's pithy witticism: "The trouble with weather forecasting is that it's right too often for us to ignore it and wrong too often for us to rely on it."[4]

More specialized metrics of weather-prediction improvement are easy to find, particularly when it comes to extreme events. We have seen that such events may be increasing in both severity and frequency and that with a rising human population we can anticipate more and more people being adversely affected by severe weather. Consequently, much effort is being put into researching severe weather with a view to improving forecasts. This emphasis is yielding results. For example, the accuracy with which we can predict the path of hurricane-eye location is better than it was. In the 1970s and 1990s, the average error in predicting the eye position 24 hours into the future was 120 and 80 nautical miles, respectively; 48 hours ahead, it was 250 and 150 nautical miles; and 72 hours ahead, it was 380 and 240 nautical miles. The general trend has been well expressed by Louis Uccellini of the National Weather Service: "We can now predict extreme weather events five to seven days in advance [compared with one day, 20 years ago]."[5]

In 1987, residents in Tornado Alley got three and a half minutes' notice, on average, that a tornado would arrive on their front door step. Currently, they get 14 minutes' notice; the goal of researchers is that this figure will increase to one hour by 2020. The EF5 tornado in Joplin, Missouri, on May 22, 2011, killed 158 people and caused $2.8 billion in damage; a future tornado of the same strength in the same unfortunate town may do just as much economic damage, but the increased warning time should reduce the number of fatalities.

Hurricane Sandy (October 22–November 2, 2012) led to 233 fatalities in eight countries, and hit the U.S. eastern seaboard badly—the storm surge in New York cut power and flooded streets and subway stations. Apart from being a disaster for the people affected, Sandy proved to be an embarrassment for the National Weather Service. Its Global Forecast System (GFS) computer model is considered to be one of the top two in the world, along with the European Centre for Medium-Range Weather Forecasts (ECMWF) model. Unfortunately, the NWS model predicted that the path of Sandy north from the Caribbean would continue out to sea, whereas it took a left turn and slammed into the U.S. coast, as the European model had predicted (figure 10.1). This failure and the severity of Hurricane Sandy has led to legislation that will provide funding for massive upgrades to the NWS model capabilities.[6]

Such errors can perhaps be put down to the statistical nature of weather prediction (any given model will never be 100% right) and perhaps to a relative underfunding of meteorological research in the

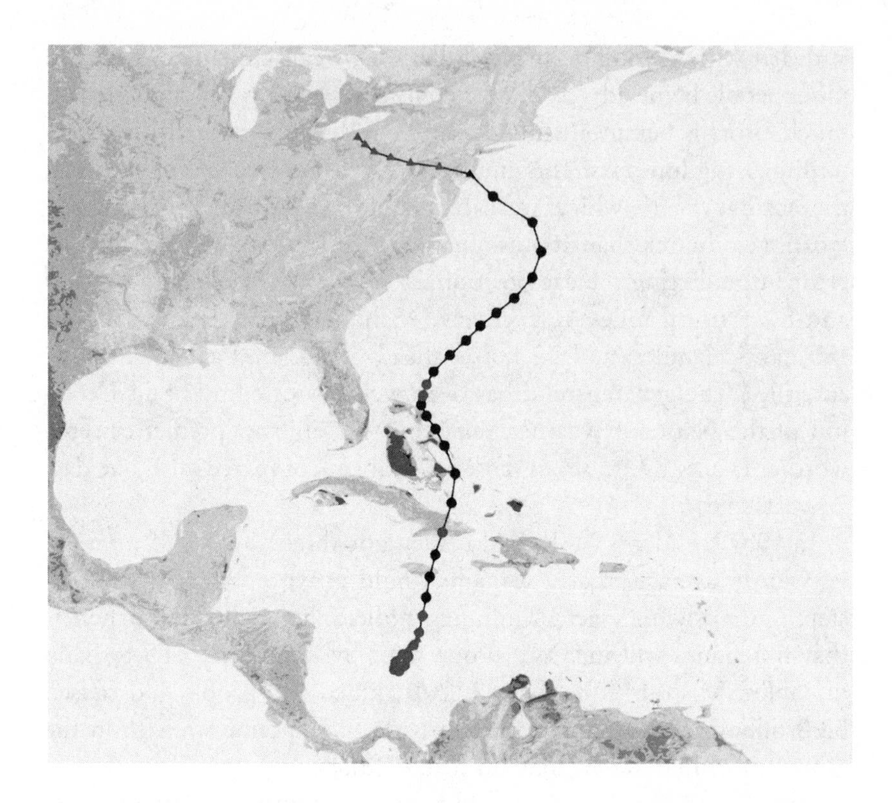

Figure 10.1 The track of Hurricane Sandy, October 22–31, 2012, from the Caribbean to the East Coast of the United States. The National Weather Service did not foresee that Sandy would hang a left. (Adapted from an image in National Weather Service, "Hurricane Sandy: October 29, 2012," National Oceanic and Atmospheric Administration, http://www.weather.gov/okx/HurricaneSandy)

United States over recent years compared with that in Europe. The trend is clear, however: forecasts are improving and will continue to improve. Why such confidence in the future? What is being put in place today (apart from more research dollars) to justify this claim? All three of the reasons listed earlier will apply in spades:

1. Knowledge of basic science has improved, so that today the fundamental physics—the notoriously difficult Navier–Stokes equations—are used extensively for forecasting. Until the 1960s, predictions were made by analogs (matching current weather conditions to past ones),

which did not work very well because of sensitivity to initial conditions (see chapter 6). Also, today's weather models apply sophisticated stochastic methods to deal with error-prone input parameters.

2. Basic input data will be denser (more measurements on Earth's surface and in the atmosphere) due to, for example, new low-Earth-orbit satellites capable of producing information in three dimensions that is updated every five minutes (currently 15 minutes).

3. More, faster, and larger computers will be able to number-crunch using smaller cells (see chapter 5), and so will capture small-scale effects such as clouds or mountains that we currently cannot model. This leads to more accurate large-scale predictions.

Better input data will include improved estimates of atmospheric temperature and wind speed aloft as a result of improved remote sensors such as weather radar and acoustic sounders. This last point highlights the nature of atmospheric physics and our understanding of it. Chaotic effects abound in this field, we now know, and so small-scale events can have large-scale consequences. For example, small-scale convective features such as thunderstorms are being predicted with increasing accuracy these days because of the increasing computer processing power available, which permits smaller computer-model cell sizes—small enough to "see" thunderstorms. Our improving knowledge of the basic physics that happens on the microscale in the atmosphere means that we can (or will soon be able to) better model large atmospheric phenomena, such as hurricanes.[7]

The Weather Industry

There are several forecasting methods that have been adopted in the past, according to circumstances. *Persistence* is a method (it barely justifies the name) that is simple in the extreme: the weather tomorrow is going to be the same as it is today. This notion works 70% of the time, in many but not all parts of the world. In a sense, it is soundly grounded in the basic science: the time scales of meteorological phenomena can be short but are usually longer than a day. The *trends* method is only slightly more sophisticated: it looks at the weather today and in the past few days and sees which way it is heading. This method requires much more data than does persistence, and requires some numerical capabilities to extrapolate current and past weather conditions into the near future. Forecasting by trends makes use of the state of the weather, today and in the recent past, and applies in a simple way the manner in which this state is changing.[8]

The *climatology* method takes a different approach, by assuming that the weather on a given day is going to be pretty close to the average. Using this approach, we can predict that in Seattle next November 15, the daily high temperature will be 52°F and there will be 0.2 inch of rain. Tucson will reach 73°F and get 0.02 inch of rain on the same day. Assuming that actual weather is not going to be too different from the average works better in some parts of the world than in others.

We have already met the analog method of weather forecasting. We note the current state of the weather and search past records for similar states in our location, and then predict that the weather tomorrow will be like it was back then. The analog method was the best available back in the day before heavy-duty computers could crunch the numbers in real time, but it was never very good because of the sensitivity of weather phenomena to initial conditions—analog predictions may work for a short peek into tomorrow, but not for long gazes into next week.

Finally, we come to the prevalent method used today for weather forecasting: it is numerical weather prediction, of course, which is based on a deep knowledge of the underlying physics and on an increasing capability to solve the mathematical equations that arise from this knowledge—and on the input data that these equations feed on. Having said that NWP is the overwhelming choice of meteorologists, we should note that there is still a role for human intervention in forecasting because people are still better than computers at pattern recognition. We have seen that different computer models can lead to different predictions for future weather, and any one model can lead to different predictions if the input data are changed slightly. Ensemble forecasting averages these predictions in a sensible way, but even this may not be as good as a combination of NWP plus experienced meteorologists when the current state of the weather is an outlier—in an atypical condition. In such a situation, it pays to employ a forecaster who knows the strengths and weaknesses of the different NWP models and who is aware of what local people have to know about their weather.

Once all the weather data have been gathered, assimilated, and put onto charts and into NWP programs, the meteorologists get down to producing global, regional, and local forecasts. As we have just seen, this is not simply a matter of running a computer program and reading the output. In practice, prediction involves most of the forecasting methods plus a little seat-of-the-pants experience from grizzled veterans. Charted input data make it easier for meteorologists to appraise the situation at a glance—they recognize patterns and physical processes (for example, a cold front on an isotherm chart). They also note from the data certain combinations of parameters, which may tell them that specific weather features may be in our immediate future. Perhaps by comparing the current data with past data, they can say that "a thunderstorm is on its way—we've seen this pattern many times before." They recognize trends and changes in the data. Data are interpolated so that local forecasts can be made for locations that are in between sites where data were gathered. NWP model outputs are compared, weighted, and combined. NWP model outputs include basic weather charts and worded forecasts (yes, computers are beginning to write the reports). At this point the grizzled veterans take over.

Nowcasting is a term used in meteorology to describe very short-range predictions—within the next six hours. Nowcasting is not so much a method as a specialized subsection of NWP. A convergence of recent technical developments will soon lead to a "major jump in nowcasting capabilities." The purpose of nowcasting is to provide up-to-the-minute highly accurate predictions of local weather—for example, during periods of severe weather or for times and locations when the weather is of particular concern (such as immediately before an important sports event).[9]

Other weather-prediction specializations deal with particular types of weather. Thus there are computer programs that look for tornadic conditions in supercell thunderstorms, and special rule-of-thumb techniques for snowfall forecasting.[10] It may seem a little unscientific to us nonspecialists for experienced meteorologists to resort to rules of thumb, but in fact this way of doing things is common in science and is perfectly valid. These techniques either are empirical or are rough approximations of physics that is currently beyond detailed calculation. In other words, an experienced meteorologist knows that a certain method works based on what she has seen before, but cannot prove it because the theory has not yet been developed or, more likely, cannot be verified because the calculations are too difficult. Examples of rule-of-thumb lore in weather forecasting include

1. Observations are more important than model predictions for forecasting the next 18 hours; for longer-range forecasts, the models are better.
2. If the relative humidity (RH) at an altitude corresponding to 700 millibars atmospheric pressure exceeds 70%, then clouds will likely develop. So imagine a contour plot of RH at 700 millibars: if the sky is currently clear, the forecast will be for clouds within the 70% RH line.
3. If a 700-millibar chart shows regions with RH exceeding 90%, rain will develop there.
4. On an 850-millibar chart, precipitation will be rain south of the –5°C (23°F) isotherm, and snow north of that line (in the Northern Hemisphere).

In a nutshell, modern weather forecasting is based on detailed mathematical calculations (NWP), perhaps averaged to account for model differences and chaotic effects, plus myriad rules of thumb that are tacked onto the model output, plus tweaks thrown in by a meteorologist to allow for local conditions (box 10.1). In years to come, we might imagine that increased computing power would reduce the human contribution to forecasting, but will not be able to entirely eliminate it.[11]

Box 10.1
Interactive Forecast Preparation System

The Interactive Forecast Preparation System (IFPS) is a software suite that, since its full deployment in 2003, has enabled National Weather Service forecasters to construct graphical representations of projected weather variables (temperature, pressure, humidity) on large-scale charts.* Various smart tools permit the forecaster to then apply NWP model output to higher-resolution grids (as small as 2 kilometers [1.24 miles]) and make subjective alterations using a graphical editor. Gridded local forecasts can be converted from pictures (good for television) into words using an automatic text formatter. The digital graphics are downloaded to customers in formats suitable for several platforms (such as hand-held devices with GPS), and the words are provided in several languages. The face of weather forecasting has come a long way since the chalkboards of the 1960s.

*See, for example, D. P. Ruth, "Interactive Forecast Preparation—The Future Has Come," in *Proceedings of the Interactive Symposium on the Advanced Weather Interactive Processing System (AWIPS)*, Orlando, Fla., January 13–17, 2002, American Meteorological Society, 20–22.

The Face of Weather Forecasting

Of all the sciences, meteorology invests by far the most effort in presenting the results of its deliberations to a nonspecialized audience—the general public.[12] Of course, the reason for this effort is that the public wants to know the results of meteorological calculations much more than it wants to know the latest results from the world of chemistry or economics, or from other branches of physics. The presentation of weather to the general public is often by way of governmental organizations, such as the National Weather Service in the United States and the Met Office in the United Kingdom. Traditionally, the forecast is provided in daily newspapers (the first such was in the *Times* of London, on August 1, 1861),[13] and later on the radio and, especially, television. Television is very well suited to the presentation of weather predictions because they are conveyed effectively by means of pictures—weather maps and satellite images. Increasingly, today's weather is told to us by other media, such as the Internet and social networks, and by private companies such as Accuweather, the Weather Network in Canada, the Weather Channel in the United States, and WeatherData. These private companies serve the increasing need for specialized weather reports, providing more detailed information that would not be of interest to the general public but that is of considerable interest to sections of industry or agriculture (soil-moisture levels for farmers, for example).

Thus, these days, a car-lot owner may be forewarned of an approaching hailstorm on his smartphone (figure 10.2), or the organizers of a

Figure 10.2 This is what baseball-size hailstones can do to a car. This hailstorm pummeled Lubbock, Texas, at the southern part of the High Plains. (From National Weather Service, "Thunderstorms Cause Wind and Hail Damage Across Southern Lubbock County: 29 April 2012," National Oceanic and Atmospheric Administration, http://www.srh.noaa.gov/lub/?n=events-2012-20120429-storms)

major sporting event or outdoor ceremony might receive detailed and frequent updates about the likelihood of rain or snow this afternoon. The crew of a cargo ship will be kept informed about the changing sea state in front of them, and construction crews working on a skyscraper will down tools and sit out tomorrow's predicted windstorm. Windsurfers can go online to find which beaches will be getting onshore winds this weekend, and the height of waves they will encounter. Ski resorts can learn about the amount and type of snow that will fall, of the likelihood of avalanches, and of temperature forecasts and windchill factors. Power companies can receive frequent updates on freezing rain (icing can bring down power lines) and lightning forecasts.

At the time of writing, 30% of Americans receive their weather information from local television, 20% from cable television, 20% from their local radio station, 20% from the web, and 10% from newspapers.

What do people want to know? In decreasing order of perceived importance: precipitation levels, temperatures, wind speeds, and humidity. Why do they want to learn these things? They want to know how to dress themselves and their children for the day ahead; to plan for the upcoming weekend, yard work, or vacation travel; to decide how and when to travel to work or school; and to schedule job activities.

The forecasts that are presented to us cover different timescales with different confidence levels. Short-range forecasts (within the next 48 hours) are more likely to be right than are extended forecasts (more than three days). The latter consist of medium-range forecasts (three to seven days) and long-range forecasts (more than seven days). Needless to say, the long-range forecasts are the most likely ones to be wrong. The weather predictions presented to us cover different scales of length as well as time: we are foretold about heat waves, which can be synoptic-scale events (of the order 1,600 kilometers [1,000 miles]); about thunderstorms, which are mesoscale events (160 kilometers [100 miles]); and about katabatic or drainage winds, such as the Santa Ana winds in southern California, which are microscale events (16 kilometers [10 miles]) resulting from dense air flowing downhill through valleys. These local, microscale events are problematic for weather forecasters because they are often on a length scale that is smaller than a weather cell used in the NWP computer programs. Public-sector weather forecasts that are handed out for distribution are often reviewed by meteorologists prior to broadcast, as we have seen, so that microscale adjustments can be made to account for local geographical conditions such as a lake or mountain.

Presentation of the uncertainty in a weather prediction is not an easy task because the general public has only an imprecise understanding of statistics. People do appreciate being told of meteorologists' uncertainties, however. Thus opinion polls show that Jo Average prefers a weather forecaster to say, "Maximum temperature tomorrow will be in the range 73 to 77°F," than to say, "Maximum temperature tomorrow will be 75°F." People want certainty but are realistic enough to know that weather is intrinsically uncertain. The credibility of a weather forecaster is not hurt when he or she says, "We don't know," but it can be ruined if he or she says, "The weather tomorrow will be calm" when there is in fact a severe storm.[14]

The most familiar of the measures of uncertainty that are presented to us is the probability of precipitation (POP), which we met in chapter 6. POP has been a feature of weather forecasts for four decades; it is popular, though most people are unsure about exactly what it means. Thus when your local forecast says that there is "a 40% POP" for tomorrow morning, the general public interprets this figure in four different ways, only one of which is correct:

1. It will rain in 40% of the region.
2. It will rain for 40% of the time.
3. It will rain on 40% of days predicted.
4. Forty percent of forecasters believe that it will rain.

If you chose the third option, then you are in company with the 19% of Americans who understand POP. Even television presentations get it wrong: my local television weather presenter may tell me, for example, that "the POP tomorrow is 40%" while displaying a graphic that shows a 40% POP for tomorrow morning and a 40% POP for tomorrow afternoon. These figures are inconsistent: if the POPs for tomorrow morning and afternoon are both 40%, then the POP for the whole day is 64%.

There is a systematic bias in the reporting of POP by television weather stations, research shows: weather presenters intentionally report higher probabilities for rain than they believe. The Weather Channel admits that when its models yield a low POP, the station boosts it a bit before presenting it to the public—say, 20% instead of 5%. Many television stations do the same, only more so. Why? The idea is that the general public will blame them if it rains when the prediction was for no rain (people often equate very low POP with no rain), whereas if the prediction is wrong the other way around—high POP but no rain—then people are relieved and do not blame the forecasters. This bias is well enough established to have its own name: the *wet bias*.[15]

The sophistication of meteorology's interface with the general public is epitomized by your local television weather presenter. Other sciences may have individuals who personify their particular field in the public domain,[16] but by comparison with this squad, the television presenters of meteorological predictions are an army. They train in colleges and universities in the various aspects of their field—media presentation with some meteorology thrown in, usually. Tom Brown, a weather reporter in Regina, Saskatchewan, once said "You don't need a [meteorology] degree to qualify as a weathercaster. But you do have to be able to explain the weather in a way that won't make real meteorologists cringe."[17] Certain schools offer training tailored for weather presenters, whereas others offer uncoordinated classes in meteorology, journalism, and so on. According to the World Meteorological Organization, weather presenters on television need a good knowledge of atmospheric physics and chemistry, plus journalism and mass-media communications.

Needless to say, over recent decades the technology of weather presentation on television has proceeded in leaps and bounds. I recall a well-known presenter in Atlantic Canada in the 1970s, Art Gould, who was famous for his memory. In those days, the weather map consisted of a chalkboard, on which Art would write from memory dozens of temperatures of all major cities across Canada and the United States while maintaining a flawless monologue. In the digital age, different skills are required. Your friendly local-television weather presenter, most people know, does not stand in front of a chalkboard weather map but instead stands in front of a blank green screen that permits the overlay of a weather map or satellite image on your digital-television screen; the presenter should not wear green or would appear partly invisible.[18] He or she knows where to point by looking at television monitors at the sides of the screen, out of view of the camera. Sometimes a giant television screen replaces the green screen, making the presenters' task easier.

Climate for Change

Humanity is learning how, by utilizing knowledge of weather mechanisms, to actually change the weather and perhaps even the climate. That is, we can manhandle the weather—physically change it to be (within limitations) more like what we want it to be. In particular, we already have some capability to make rain and are developing other possible methods of weather modification, and we may be able to reduce the global mean temperature (as well as raise it). The climate part of this capability goes by the name of *geoengineering* and is new;

the weather part—at least the rainmaking—is older. Both are controversial, in part because they just might work.

Here is the problem. We don't really know the downstream risks for weather modification or climate change—recall the extreme sensitivity to parameter change of chaotic systems—and so we cannot say what the long-term consequences may be for seeding a cloud in Wyoming today or increasing the atmospheric albedo over Antarctica tomorrow. Some people regard our increasing capabilities in this field as the answer to our weather and climate problems, while others see them as tickling the dragon's tail. In the words of one expert: "Personally I find this stuff terrifying but we have to compare it to doing nothing, to business-as-usual leading to a world with a 4°C rise."[19]

You may be tempted to think that an average global temperature increase of 4°C (7°F) is not a bad thing—after all, it is not all that much and, in our part of the world at least, it could mean no more than the difference between wearing a long-sleeved shirt and a T-shirt. But, in fact, such a rise would lead to very serious consequences for global weather and climate. This temperature increase represents extra energy being put into our atmosphere by the sun. We have seen how dynamic the atmosphere–ocean system is—an increase in energy makes it more so. The atmosphere–ocean system being dynamic, a change in climate automatically means a change in weather. The net effect may be an average increase in temperature of 4°C, but the effect on different parts of the world will be much greater than this. Some regions may cool, while others may heat up a lot. Higher average temperature means increased water vapor in the atmosphere and more rainfall—though it will not be uniform. The global circulation patterns will get stoked up; whether they remain stable at the higher temperatures depends on the temperature rise and on the climate model—a rise of 2°C (3.6°F) may be tolerable, current models concur, but beyond that we are in the lap of the gods, with the likelihood of positive-feedback effects kicking in that can take the climate into a new state, far from the present zone of stability. The general prediction is for more extremes as average atmospheric temperature rises: wet places become wetter, with rainstorms dumping more water than they did before the rise. Dry places become drier, with droughts lasting longer. People living in subtropical regions are likely to feel the changes first, and most. There will be more extreme cyclones. Different nations will experience the effects differently, depending on geography. Thus the melting of polar land ice (glaciers and ice sheets) plus thermal expansion will cause sea levels to rise, greatly affecting low-lying countries such as Bangladesh and the Seychelles, affecting countries with coastal cities somewhat less, and affecting landlocked countries such as Switzerland not at all—or, more accurately, not directly. The regional differences in climate change will doubtless lead to political difficulties between nations; some pundits foresee these difficulties escalating into wars, as migrants from drought-stricken regions pour across borders, as we speculated earlier.[20] All this from a small rise in average temperature.

So in what ways are we able to modify weather and geoengineer climate? Let us consider weather modification first. By far the best developed and best known manmade modification is rainmaking, induced by the seeding of clouds (figure 10.3). Rainmaking can relieve a drought and clear the atmosphere of pollutants. Vincent Schaefer first tried this idea in 1947 by flying over the Adirondacks and releasing dry ice into a cloud. Snow began to fall from the cloud base, along a track following the plane. Although the snow evaporated before reaching the surface, the principle had been established. Since then, silver iodide has become a more popular seed; it has a crystal structure resembling that of ice, and

Figure 10.3 Cloud seeding, from the ground and from the air. The seed (silver iodide, dry ice, seawater droplets) acts as condensation nuclei, converting water vapor in the cloud into liquid water that falls as rain. (Thanks to DooFi for this line drawing)

so small crystals of silver iodide can act as condensation nuclei for water vapor in clouds, causing water-droplet formation and, hopefully, precipitation. The size of a condensation nucleus is much less than that of the water droplet it gives rise to, and so the mass of rain that can be produced by seeding a cloud is much greater than the mass of seeding material.

Cloud seeding has matured since Schaefer's initial efforts. As well as using aircraft, we can seed clouds using rocket launchers and anti-aircraft guns, and may in the future use drones, which can seed a thundercloud base (which manned aircraft cannot do, for safety reasons). Results of seeding programs have nevertheless been indifferent. Critics say that the technique cannot really generate rain; all it can do is accelerate the production of precipitation that was going to happen anyway.

The second strand of weather modification—in its early stages at the time of writing—is storm prevention or reduction. Several ideas have been proposed to mitigate the effects of severe storms:

1. Apply laser beams to discharge lightning in storms that are building up.
2. Coat the surface of warm seas beneath developing hurricanes to prevent droplet formation (robbing the storm of energy).
3. Seed the eye wall of hurricanes (to disperse energy).
4. Deploy soot to the outer walls of hurricanes to absorb sunlight and alter convection currents.
5. Disperse hygroscopic gel into storms (returning water to the surface).

Some of these ideas are realizable for small storms, such as thunderstorms, but will be more difficult to implement on the synoptic scale. Thus huge amounts of soot would have to be dispersed to effect development of hurricanes.

For decades regarded as at best a fringe science, weather modification—to relieve drought or reduce the ferocity of storms—is now at the forefront of meteorological research.[21] China has the largest such research program, with 1,500 weather-modification professionals, plus 30 aircraft and crews, 37,000 part-time workers (read peasant farmers), 7,113 anti-aircraft guns, and 4,991 rocket launchers. If this description makes weather modification sound like a military campaign, that is not inappropriate given the military interest that exists in researching weather-modification techniques as force multipliers.[22]

Geoengineering, if realized, would operate on a much larger scale and act over much longer timescales, making weather-modification efforts seem puny by comparison. The (scary to some, as we have seen) potential climate-altering techniques that have been proposed and simulated by experts include

- Aircraft spraying sulfur into the stratosphere to mimic volcanic eruptions
- Artificial trees to remove carbon dioxide from the atmosphere
- Ships spraying dense plumes of particles into the atmosphere to alter oceanic clouds
- Sulfate aerosols injected into Arctic air to reverse sea-ice melting
- Dumping iron into the seas to trap carbon and stimulate plankton blooms

A pictorial summary of proposed geoengineering techniques appears in figure 10.4. These ideas have been virtually tested—input to climate models to see their effects and any medium-term unintended consequences. Here, of course, is the rub. We know, because the atmosphere–ocean system is chaotic, that we cannot foresee the long-term consequences of our actions and so cannot be sure that the (largely irreversible) geoengineering solutions are not in fact geoengineering problems. Modeling suggests, for example, that injecting aerosols into Arctic air would indeed reduce temperatures in that region and so slow down or reverse the melting of ice sheets, but it would also alter global rainfall patterns and profoundly disrupt the monsoons on which India depends. It would also result in a complete drying of the Sahel region of Africa. These downstream effects are not obvious, and that is why the whole idea of geoengineering is terrifying to some climatologists.[23]

***

The accuracy of weather predictions is increasing and will continue to increase as understanding, data quality and quantity, and computing power all get better with each passing decade. The dominant prediction method today is a combination of numerical weather modeling coupled with tweaks from experienced meteorologists to account for

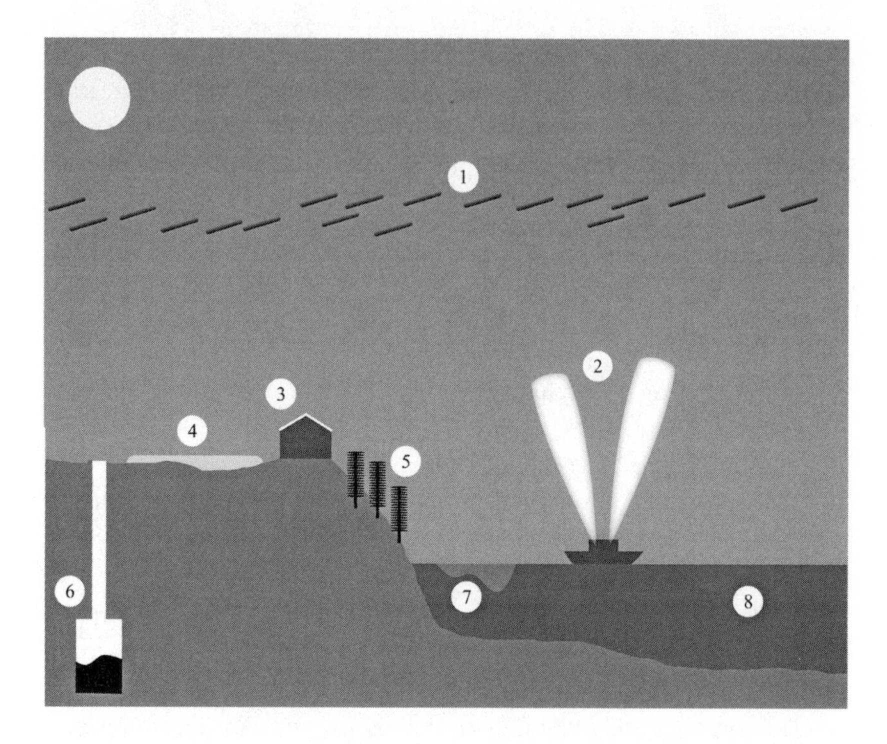

Figure 10.4 The spectrum of climate-engineering proposals, from harebrained to feasible: (*1*) positioning trillions of tiny aluminum mirrors in space, to deflect sunlight; (*2*) seeding clouds with seawater, to cause rain; (*3*) constructing pale roofs to reflect sunlight; (*4*) growing pale drought-resistant crops, to reflect sunlight; (*5*) erecting artificial trees, to capture atmospheric carbon dioxide and bury it; (*6*) burning and then burying agricultural carbon waste; (*7*) adding iron to oceans to encourage phytoplankton to sequester atmospheric carbon; (*8*) creating and deploying synthetic microbes to sequester atmospheric carbon.

microscale conditions. Weather presentation is an increasingly sophisticated industry. Cloud seeding to induce precipitation has been around for 60 years, but its efficacy is debatable. Storm mitigation and the geoengineering of climate are being researched, though their possible long-term consequences concern most researchers.

And That Wraps Up Your Weather for Today

Prediction is very difficult, especially about the future.

Niels Bohr

Our examination of weather and climate science is now finished, if not complete—completeness, in a few hundred pages, is not possible for these complex subjects. I hope that you have learned something of the intricacy and depth of atmospheric and oceanic physics and grasped some of the underlying, overriding principles. My own journey has left me with a great appreciation of the entire weather-predicting enterprise as an achievement that we must include among the greatest that our species has created. The intellectual effort that has been expended over the past century or so may be matched by other scientific constructs—quantum theory or evolutionary theory, perhaps—but the material effort and organization that supports it is unrivaled. The rapid and impressive development of, say, pharmaceuticals requires extensive trials and much data gathering, but nothing like the day-to-day data acquisition needed to support the weather-prediction industry. Perhaps the Manhattan Project in World War II—developing the atomic bomb—exhibited something of the intellectual and physical effort, but even this project is dwarfed by the larger and much longer-running international efforts to understand the global weather and climate system.

Here are a few more stray thoughts about weather and climate physics, to round off this book.

***

The photograph was taken at 1:05 P.M. on June 23, 2015, facing west—to the open Pacific—from a beach near Tofino on the west coast of Vancouver Island (figure C.1). The weather at this latitude (49.1°N) comes out of the west, most of the time, as you will appreciate from our discussion of the Coriolis force and trade winds. For the previous several days, the weather had been fine—warm and sunny, as in the picture—but the forecast was for change: two days of cloud and light rain, followed by a return to warm and sunny conditions. It turned out just as forecast.

This prediction impresses, especially given that the weather is appearing from a void—the nearest population centers due west of Tofino are thousands of miles away in Japan. Yet satellite and other data are sufficient for numerical weather prediction models to predict the weather in this small Canadian community accurately. More than this, though: the knowledge of meteorologists is deep enough for them to read skies

Figure C.1 A view westward from Schooner Beach, near Tofino, on the west coast of Vancouver Island, British Columbia, Canada. There is a story in the clouds. (Photo by the author)

and estimate likely weather even without the sophisticated world-wide data-gathering and numerical-forecasting infrastructure. I showed this picture to NOAA meteorologist Chris Wamsley and asked him if the sky contained any hints of the weather to come; his response is worth quoting: "Anytime I see high wispy clouds, I think of remnants from strong storms in the past and being carried and eroding at the same time from previous storms upstream or mainly from the west. This would be a sign [that] storm activity has occurred recently from the west and possibly rain soon to come in the next day or two." Chris went on to say that such (cirro-form) clouds won't predict all storms because some develop directly overhead, but is a good indicator, as we saw in chapter 7.

We have seen that modern weather forecasting consists of numerical predictions tweaked by input from experienced meteorologists. Both rely on the known physics of weather systems: the computer models have these laws of physics built in and the meteorologists have learned, digested, and applied such laws enough times to develop rules of thumb—intuition—about the way that the weather world works.

***

Tell them what you are going to tell them, then tell them, then tell them what you have told them. This mantra from chapter 8 has been a feature of the text in that you have been exposed to the same aspects of our subject again and again from different perspectives, to emphasize a difficult or an important point and to increase the chance that, like incoming solar radiation, it is absorbed. Of the dozen popular science books that I have written, on all kinds of topics from sailing ships to biophysics, this one has covered the most varied and most complex physics. As a consequence, I have given quite a great deal of thought to the best method of presentation—slightly technical, to provide some depth; not much math, to aid readability; and repetition from different viewpoints, to increase learning without being too boring. These repetitions arose naturally because of the interdependence of different physical elements of meteorology and climatology. Thus, for example, the heating of oceans was introduced in chapter 2, revisited in chapter 8, and mentioned in all the other chapters except chapter 6—oceans absorbing solar radiation, oceans interacting with the atmosphere, oceans spawning tropical cyclones, oceans affecting climate change. Weather extends from the atmosphere to the top layer of the oceans, and climate extends all the way to the bottom.

***

Statistical fluctuations can't be predicted, but the averages can—that was one of the outcomes of chapter 6. So does this observation mean that climate (average weather) is predictable? In some ways, it is more predictable than weather, but the difference is only a matter of degree and not kind. We have seen that there are chaotic processes in climatology—unsurprising in so complex and nonlinear a field. Hence predictability is limited. More important for long-term climate-model predictions, there are natural events that cannot be modeled and that significantly influence climate. We know that volcano eruptions will occur but have no theory that enables us to say how many will spew how much ash into the stratosphere in the 2060s, for example. Climatology is influenced by what goes on underground as well as what goes on under the ocean surface. Again, the realm of climatology is deeper than that of meteorology.

And broader. If human activity is significant enough to modify the climate of our planet, as increasingly seems to be the case,[1] then to predict climate development, we must predict the development of world economics. (Will developing nations follow our own fossil-fuel road to economic development? When or to what extent will renewable and clean fuel sources become economically viable? How much fuel will humanity consume in 50 years?) Economics is infamously unpredictable and poorly understood—witness the financial crisis of 2008—and consequently the extent of our General Circulation Model accuracy is not set by our limited understanding of the underlying physics, but by our limited understanding of ourselves.

Appendix

This appendix presents technical calculations referred to in the main text. I use metric units throughout, and the absolute (Kelvin) temperature scale.

Electromagnetic Radiation from the Sun

The total electromagnetic power emitted by our sun is an awesome 3.846×10^{26} watts. The sun's radius is 696,000 kilometers, and so the power density at the surface of the sun is 63 MW m^{-2}. Our Earth has a mean radius of 6,371 kilometers and is 150 million kilometers away from the sun (the geometry is sketched in figure 1.1*a*), so the mean solar power reaching Earth is 1.73×10^{17} watts and the mean power density bathing Earth's disk amounts to 1,365 W m^{-2}.

Stefan–Boltzmann Law

The emissive power of the sun (the power density emitted per square meter of surface) is a simple function of the sun's surface temperature, T_S:

$$p_s = \sigma T_S^4$$

where temperature is in Kelvin—that is, degrees absolute. This is the *Stefan–Boltzmann Law,* derived from thermodynamic principles in the nineteenth century. The constant σ is the Stefan–Boltzmann constant and has a value of $\sigma = 5.67 \times 10^{-8}$ W m^{-2} K^{-4}. We calculated the power density of EM radiation emitted at the surface of the sun in the last paragraph, and so can determine the solar surface temperature: about 5,760 K. From the value we obtained for solar power bathing Earth, we see that the power density that impinges on our planet is about 1,365 W m^{-2}. Divide by a factor of 4 for geometrical reasons made plain in figure 2.1, and we obtain a mean incident solar power on the surface of planet Earth of about 341 W m^{-2}. Deduct 30% to account for the fraction that is reflected back into space, and apply the Stefan–Boltzmann equation to obtain a surface temperature for Earth of T_E = 254 K. This simple calculation is based on a number of assumptions, the main one being that both Earth and the sun are perfect blackbodies.

From thermodynamic principles, the emission spectral irradiance of a blackbody (the power emitted per unit wavelength per unit solid angle) is a function of both wavelength and temperature:

$$B(\lambda, T) = \frac{2hc^2}{\lambda^5} \frac{1}{\exp(hc / \lambda k_B T) - 1}$$

where h, c, and k_B are basic physical constants of known value (for example, c is the speed of light, which is 2.998×10^8 m s^{-1}). This formula is known as the *Planck equation.* Substitute for the surface temperature of the sun and of Earth, and we obtain the power spectra plotted in figure 1.2. Integrating the Planck equation over solid angle and wavelength yields the Stefan–Boltzmann equation.

Cloche Calculations

From figure 4.1, we see how a single pane of glass increases the temperature inside a cloche, as follows. The Stefan–Boltzmann Law tells us how the equilibrium blackbody temperature T_0 of the surface inside the cloche (initially assuming no glass [see figure 4.1*a*]) is related to incoming solar power P_0, that is,

$$P_0 = \sigma T_0^4$$

In other words, the incoming shortwave density of power that is emitted by solar radiation equals the outgoing longwave density of power that is emitted by the surface. Let us say that the power density that reaches the surface is such that the ambient temperature is T_0 = 270 K. Now add the glass, so that the energy flow is as shown in figure 4.1*b*. Balancing the budget, we see that outside the cloche

$$P_0 = \sigma T_1^4$$

so that $T_1 = T_0$. Inside the cloche we see that

$$P_0 + \sigma T_1^4 = \sigma T^4$$

and so $T = 2^{1/4} T_0$, or T = 321 K, or 48°C as stated in chapter 4.

It is not difficult to generalize this calculation from 1 to n layers of glass. For two layers of glass, we find that $T = 3^{1/4} T_0$ = 355 K; for n layers, $T = n^{1/4} T_0$. Thus Venus, which has a surface temperature that is four times its blackbody temperature ($T = 4T_0$), can be thought of as a greenhouse with not one but 256 layers of glass.

Thin Atmosphere Power Budget

For the thin atmosphere planet of figure 4.3, we can carry out a similar calculation. In space, the energy-balance equation is

$$P_0 = (1 - A)\sigma T_E^4 + \frac{1}{2}\sigma T_A^4$$

In the atmosphere, it is

$$aP_0 + A\sigma T_E^4 = \sigma T_a^4$$

On the planet surface, it is

$$(1-a)P_0 + \frac{1}{2}\sigma T_a^4 = \sigma T_E^4$$

Here $T_{E,a}$ refer to the temperature of Earth's surface and atmosphere, and a, A refer to the fraction of shortwave and longwave radiation that is absorbed by the atmosphere, as shown in figure 4.3. From the first and last of these equations, we obtain the following expression for surface temperature:

$$T_E = \left(\frac{2-a}{2-A}\right)^{1/4} T_0$$

where T_0 is the blackbody temperature of the planet if it had no atmosphere, $T_0 = (P_0/\sigma)^{1/4}$. Because a, A lie between 0 and 1, the surface temperature can vary between $0.84T_0$ and $1.19T_0$, as stated in chapter 4.

Earth's Power Budget

From figure 4.4, the energy or power balance equations are, for the atmosphere,

$$aP_0 + A\sigma T_E^4 = \sigma T^4 + \sigma T'^4$$

and for the surface,

$$(1-r-a)P_0 + \sigma T^4 = \sigma T_E^4$$

Assuming the values for parameters a, A, and r given in chapter 4, and assuming the observed Earth surface temperature of $T_E = 288$ K, we find that the lower atmosphere is at temperature $T = 254$ K and the upper atmosphere at temperature $T' = 243$ K. Thus the power of back radiation from the atmosphere exceeds the power it radiates to space, by 19% in this simple model.

Furthermore, if we assume the observed atmospheric lapse rate near the surface—that is, 6.5 K km^{-1}—we see that the lower atmosphere effective height is 5.4 kilometers, corresponding to the middle of the troposphere. The effective height of the top of the atmosphere is at 8.6 kilometers, not far from the tropopause.

Visibility Through Watery Air

Say the density of liquid water in the air is w g m^{-3}, and water droplet diameter is d. Then the mass of one spherical droplet is $m = \pi\rho d^3/6$ g, where ρ is water density. The number of droplets per cubic meter is (assuming all droplets are the same size) $n = w/m = 6w/\pi\rho d^3$. Thus, on average, there is one droplet per $1/n$ cubic meters. The probability of a ray of light intercepting a drop over a path length of $n^{-1/3}$ meters is the ratio of droplet cross section to the area occupied by one droplet over this path length—that is, $p = ¼\ \pi d^2/n^{-2/3}$. The mean free path for light traveling through air containing these water droplets is denoted λ, where $(1-p)^{\kappa}\epsilon = ½$. Here $\kappa = \lambda n^{1/3}$. For small n this equation simplifies to

$$\lambda = \frac{d}{3}\frac{\rho}{w}$$

From this equation, the numbers of box 7.1 are derived.

Hailstone Size

We show that hailstone size increases as the square of updraft air speed, as claimed in chapter 7. For simplicity, assume spherical hailstones. The mass of a hailstone of radius r is

$$m = \frac{4}{3}\pi r^3 \rho$$

where ρ is ice density. The aerodynamic drag force is

$$D = \frac{1}{2} c_D \rho_{air} \pi r^2 v^2$$

where c_D is aerodynamic drag coefficient, ρ_{air} is air density, and v is updraft speed. If the hailstone is just held up by the drag force, then $D = mg$ where g is the acceleration due to gravity and so

$$r = \frac{3}{8} c_D \frac{\rho_{air}}{\rho} \frac{v^2}{g}$$

Glossary

Advection The transfer of heat by the physical movement of hot material.

Advection fog A ground cloud that forms when warm, moist air passes over a cooler surface.

Air mass A synoptic-scale volume of air with roughly the same density and temperature throughout.

Albedo The reflectance of a body—the fraction of incident electromagnetic radiation that reflects off its surface.

Anticyclone Large-scale winds spiraling out from a high-pressure center—clockwise in the Northern Hemisphere.

Automatic Surface Observation System (ASOS) An automated weather station.

Blackbody An idealized object that absorbs all incoming electromagnetic radiation.

Blackbody radiation Electromagnetic radiation that is emitted by a blackbody that is in thermal equilibrium with its surroundings. The spectrum is described by Planck's Law.

Blocking pattern A pressure disturbance that stalls the normal movement of weather systems.

Butterfly effect Extreme sensitivity to initial condition, characteristic of a chaotic physical system.

Carbon cycle The exchange of carbon among the atmosphere, biosphere, and the land and ocean.

Cell The physical volume adopted for a data point in a computer model of weather or climate. Thus the temperature, pressure, and so on are assumed to be identical everywhere within the cell.

Chaotic A deterministic physical system that evolves in time in a manner that depends very sensitively on its starting position or state is chaotic. So imperfect knowledge of the initial state leads to unpredictable behavior that appears random.

Cirro-form High, wispy ice clouds.

Climate Weather patterns, averaged over time (usually 30 years).

Climate attribution The forensic investigation of individual weather events for telltale signs of climate change.

Cloud seeding Spraying condensation nuclei (such as silver iodide particles) into clouds to induce rain.

Cloud-to-ground lightning A discharge between cloud and ground, seen as a linear or forked track.

Condensation nucleus A microscopic particle on which water vapor can condense to form a droplet.

Conditionally stable air Air in which a saturated parcel acts as if it is in unstable air, whereas an unsaturated parcel acts as if it is in stable air. It is the most common type of atmospheric stability.

Conduction The transfer of heat by physical contact.

Conservation law In physics, a law stating that a property of a physical system has a constant value. For example, its energy may take different forms, but the sum of all forms is constant.

Coriolis force An inertial (sometimes called "fictitious") force that acts on a body that is in motion relative to a rotating reference frame. Thus an observer on Earth's surface can detect a Coriolis force that deflects objects that move over the surface. The force is small except for movements of hundreds of miles.

Cumulo-form Tall, dense fluffy clouds.

Cyclone Large-scale winds spiraling in to a low-pressure center—counterclockwise in the Northern Hemisphere.

Derecho A straight-line windstorm, associated with bands of thunderstorms.

Deterministic A deterministic physical system is one that evolves in time in a manner that is entirely predictable.

Dust devil The smallest and shortest-lived atmospheric vortex. It is common near a warm land surface beneath unstable air.

Electromagnetic (EM) energy Energy in the form of photons that are radiated by a body, an emitter. Visible light is a form of electromagnetic energy.

El Niño–Southern Oscillation (ENSO) A quasiperiodic variation of winds and of sea surface temperatures in the tropical Pacific.

Ensemble forecast A weather or climate-model forecast in which a spread of initial conditions is assumed, because exact conditions are not known. This spread means that predictions are also spread, with known probabilities.

Evapotranspiration Loss of water from soil due to evaporation and to transpiration from the leaves of plants.

Eye wall The region of maximum wind speeds near the center of a hurricane.

Faint young sun paradox The apparent contradiction between the expected freezing temperatures of the early Earth and observations indicating the presence of liquid water at that time.

Ferrel cell The average motion of mid-latitude air, descending in the subtropics and rising further poleward.

Föehn A warm, dry wind that descends the lee slope of a mountain. It is also called a chinook.

Front In meteorology, a transition zone between air masses of different densities.

General Circulation Model (GCM) Large computer simulations of Earth's climate.

Geoengineering The deployment of technological means to modify climate.

Geostationary orbit An equatorial orbit in which a satellite maintains a fixed position relative to Earth's surface.

Geostrophic A geostrophic current is an oceanic flow in which pressure gradient force and Coriolis force balance. A geostrophic wind would run parallel to isobars, though in practice the presence of friction modifies this idealized behavior to some degree.

Global conveyor belt The system of deep-ocean circulations driven by temperature and salinity differences.

Graupel Snow pellets formed when supercooled water droplets freeze onto a falling snowflake.

Greenhouse effect The heating of a planet's surface due to the trapping of solar electromagnetic radiation by the atmosphere.

Gust front A strong radial wind burst due to a thunderstorm downdraft.

Gyre A large system of wind-driven rotating ocean currents.

Hadley cell A tropical atmospheric circulation, with air rising near the equator and descending in the subtropics.

Heat capacity The ability of a substance to absorb heat.

Heat of vaporization The latent heat that is needed to convert a liquid to a gas at the same temperature.

Hindcasting Testing a weather or climate model by using it to "predict" events that have already occurred.

Hurricane A synoptic-scale vortex arising over warm oceans. It is also known as a cyclone or a typhoon.

Hysteresis A lag in the response of a physical system to a change in forces acting on it.

In-cloud lightning A discharge within a cloud that appears as an undifferentiated flash. It is also known as sheet lightning.

Infrared Electromagnetic radiation that is of longer wavelength (lower frequency) than visible light. It is detectable as heat.

Intertropical Convergence Zone (ITCZ) The belt of converging trade winds near the equator. It is also known as the doldrums.

Irradiance The power density of solar radiation at the top of the atmosphere.

Jet stream A fast-flowing narrow air current. The major jet streams are located near the tropopause.

Lapse rate The rate at which air temperature decreases with increasing altitude. The environmental lapse rate is that of stable air; the dry adiabatic lapse rate is that of a parcel of unsaturated air; the saturated adiabatic lapse rate is that of an air parcel saturated with water vapor.

Latent heat Energy that is absorbed or released by a body without changing its temperature.

Level of free convection The altitude at which a parcel of saturated air becomes warmer than its surroundings, and so rises freely.

Lifted condensation level The lowest altitude at which water vapor in a rising parcel of air begins to condense. It is often the level of a cloud base.

Longwave radiation In climate physics, the microwave part of the electromagnetic spectrum.

Luminosity The total power output of the sun.

Meddy A large eddy of water in the Atlantic Ocean, differing in temperature and salinity from its surroundings, that originates in the Mediterranean Sea.

Microwave Electromagnetic radiation of lower energy than visible light or infrared.

Milankovitch cycles Slow periodic changes in Earth's orbital parameters that result in climate changes.

Multicell thunderstorm A cluster of single-cell thunderstorms.

Negative feedback A process in which disturbances in a system tend to die out. The disturbance itself discourages disturbance—a hallmark of stability.

NEXRAD Doppler radars that measure atmospheric phenomena. The acronym stands for NEXt generation weather RADar.

Nimbo-form Rain clouds.

Nowcasting Very short term, local weather forecasting.

Numerical weather prediction (NWP) models Computer simulations of weather. Given accurate data about current conditions, NWPs can predict the state of the weather a number of days ahead.

Parameter A numerical characteristic that defines a physical system. For a simple pendulum, the parameter is its length. For more complicated systems, several parameters are needed to define it uniquely.

Parameterization The replacement of a small-scale or complex physical process in a weather or climate model with a simpler process.

Parcel A volume of air that can be regarded as having the same meteorological properties, such as temperature and humidity, throughout.

Polar cells Atmospheric circulations in which air rises at latitudes 60° to 70°N and S and descends near the poles.

Polar front The boundary between the polar cell and the Ferrel cell in each hemisphere, characterized by large temperature differences.

Polar orbit An orbit in which a satellite passes over both poles.

Positive feedback A process in which small disturbances become magnified. The disturbance generates further disturbance—a hallmark of system instability.

Power Energy per unit time—the flow of energy.

Proxy data Measurements that are indirect estimates of some desired data. Thus ice-core samples are a climate proxy for atmospheric carbon dioxide gas levels.

Radiation Spontaneously emitted energy, here electromagnetic (photons) though other forms are possible (for example, neutrinos).

Rossby waves Giant meanders in the polar jet stream that are responsible for much of the mid-latitude weather.

Shortwave radiation In climate physics, that part of the electromagnetic spectrum that includes visible light.

Single-cell thunderstorm The weakest and briefest type of thunderstorm, characterized by little or no wind shear.

Snowball Earth The hypothesis that Earth's surface was mostly frozen in the distant past.

Spectrum A range of energies of, here, electromagnetic radiation, for which energy is proportional to frequency.

Squall line A linear series of thunderstorms that forms along a cold front.

Stable air Air that is not prone to vertical movements and, hence, is layered.

Standard deviation A statistical measure of variation within a set of data points. About 68% of samples lie within one standard deviation of the mean value, or average.

Stefan–Boltzmann Law The physical law that relates the power density at the surface of a blackbody radiator to its surface temperature.

Strato-form Flat diffuse blankets of cloud.

Stratosphere The atmospheric layer immediately above the troposphere, characterized by temperature increasing with altitude.

Supercell thunderstorm The most power and long-lived type of thunderstorm, characterized by strong wind shear and potentially leading to tornadoes.

Synoptic scale Length scale of the order of 1,000 kilometers (600 miles).

Systematic error A measurement error that is not random—for example, an error due to incorrect calibration of a measuring instrument.

Temperature inversion An atmospheric condition in which temperature increases with altitude.

Thermal equilibrium Two bodies are in thermal equilibrium if heat energy is able to flow between them but doesn't.

Thermocline A thin layer of fluid that exhibits rapid temperature changes with depth. In the oceans, the thermocline is usually near the surface.

Thermodynamics The study of heat and its relationship with other forms of energy.

Thermohaline circulation Large-scale ocean circulations that arise from density gradients, due to surface heating and freshwater outflow.

Tornado A powerful column of rotating air between cumuliform clouds and the ground.

Tropopause The upper boundary of the troposphere.

Troposphere The lowest layer of the atmosphere, between the surface and the stratosphere.

Ultraviolet Electromagnetic radiation with greater energy (that is, shorter wavelength, higher frequency) than visible light.

Unstable air Air in which movement is reinforced by positive feedback. Rising air is impelled to rise faster, for example.

Water vapor The gaseous form of water.

Wet bias The tendency for weather forecasters to boost low probabilities of precipitation.

Wind shear Variation in wind velocity that exerts a turning force.

Wind stress Shear stress due to the wind—frictional force acting on, for example, the ocean surface due to the wind.

WMO World Meteorological Organization.

Notes

Author's Note

1. For the record, 1 meter = 1.09 yards, 1 kilometer = 0.62 mile, and 1 kilogram = 2.2 pounds. (Readers with a scientific background may cringe at the last equality, as it seemingly conflates mass and weight—live with it.) To convert centigrade into Fahrenheit, multiply by nine-fifths and add 32. The appendix contains a few technical calculations, all performed in metric units.

Forecast

1. I am walking you through something of a minefield here, as will become apparent in chapter 4. The phrases "short term" and "long term" have different meanings in weather forecasting and climate prediction. Also, the type of data necessary for climate prediction is more extensive than that for weather forecasting.

2. The ferocious tornado that flattened Moore, Oklahoma, in the winter of 2013/2014 was immediately replaced in the headline news by the "polar vortex" that put much of North America in the deep freeze. Before this book goes to press, more of these extreme-weather events that affect our lives will occur.

3. I feel the need to insert these words because one aspect of our subject—climate change—polarizes opinion like nothing else, except maybe gun control or abortion rights. Climate is simply weather that is averaged over space and time (it has been said that "climate is what you expect—weather is what you get"); as such, it is appropriate to incorporate the subject into this book.

4. David Derbes is Head of Science at the University of Chicago Laboratory Schools, author of physics papers and textbooks, a lauded teacher, and a good friend from postgraduate days at the University of Edinburgh. Note David's mixing of units: here, I mostly use metric with American units added parenthetically afterward. As a scientist, I naturally prefer the rational metric system, but many of the practical sciences in the United States adhere to the older units: feet, degrees Fahrenheit, calories. I am aiming for qualitative explanations in this book, not quantitative calculations, so this ordinal sin of mixing units is (I hope) forgivable.

5. The heat wave in France in 2003 killed nearly 15,000 people, mostly elderly. That year, Europe experienced its hottest summer since the Middle Ages.

1. Feeling the Heat

1. Polarity flips happen to Earth's magnetic field, but not often and certainly not regularly. In fact, the flip seems to be chaotic, happening on average every 300,000 years—and yet the last one occurred 780,000 years ago. Polarity flips of Earth's magnetic field appear to have no consequences for our climate. See, for example, "2012: Magnetic Pole Reversal Happens All the (Geologic) Time," November 30, 2011, NASA, http://www.nasa.gov/topics/earth/features/2012-poleReversal.html.

2. R. A. Goldberg, "A Review of Reported Relationships Linking Solar Variability to Weather and Climate," in *Solar Variability, Weather, and Climate*, ed. J. A. Eddy (Washington, D.C.: National Academies Press, 1982); H. Miyahara and Y. Yokoyama, "Influence of the Schwabe/Hale Solar Cycles on Climate Change During the Maunder Minimum," *Proceedings of the International Astronomical Union* S264 (2010): 427–433; and O. M. Raspopov et al., "The Influence of the de Vries (~200-year) Solar Cycle on Climate Variations: Results from the Central Asian Mountains and Their Global Link," *Paleogeography, Paleoclimatology, Paleoecology* 259 (2008): 6–16. For less technical articles on the connection between solar variation and climate, see United States Geological Survey, "The Sun and Climate," Fact Sheet FS 095-00, August 2000, https://pubs.usgs.gov/fs/fs-0095-00/fs-0095-00.pdf; and National Aeronautics and Space Administration, "Why NASA Keeps a Close Eye on the Sun's Irradiance," May 25, 2010, http://www.nasa.gov/topics/solarsystem/features/sun-brightness.html.

3. Kelvin made a number of important contributions to many fields of physics, including thermodynamics. The Kelvin scale of temperatures (also known as the absolute temperature scale) is named after him.

4. For more on the debate about the age of Earth at the turn of the nineteenth and twentieth centuries, see, for example, M. Livio, *Brilliant Blunders: From Darwin to Einstein; Colossal Mistakes by Great Scientists That Changed Our Understanding of Life and the Universe* (New York: Simon and Schuster, 2013), chap. 4; and K. Sircombe, "Rutherford's Time Bomb," *New Zealand Herald* (Auckland), May 15, 2004. Modern measurements of the total power emanating from natural radioactive sources within Earth are reported in L. Rybach, "Geothermal Sustainability," *Geo-Heat Centre Quarterly Bulletin* 28 (2007): 2–7. Note that the amount of radioactivity, and so the amount of internal heat generated by it, will have been greater in past eons as short-lived isotopes will have added their contributions before disappearing. A more recent and more precise estimate of the age of Earth is 4,568 million years. See A. Bouvier and M. Wadhwa, "The Age of the Solar System Redefined by the Oldest Pb-Pb Age of a Meteoritic Inclusion," *Nature Geoscience* 3 (2010): 637–641.

5. Grammatically, these should be two words but over the years physicists have married them into one, "blackbody," especially when used as an adjective.

6. The universe is a blackbody, and the cosmic background radiation has a spectrum that is very close to that of a blackbody at temperature 3 K in thermal equilibrium (with what? you may well ask). A black hole is, perhaps unsurprisingly, a blackbody; the blackbody radiation it emits is called *Hawking radiation*. A furnace with a small hole in it is a good practical approximation to a blackbody, and the spectrum of EM radiation that emerges from the small hole is very close to the blackbody spectrum for an object at the furnace temperature.

7. This case is common. It assumes that the rising air does not transfer heat to nearby air that it passes through—technically known as the *adiabatic* assumption. It is common in practice because air is not an efficient conductor of heat.

2. Under the Heavens and the Seas

1. The estimation of power begins with this value, but will be whittled down as the radiation penetrates the atmosphere, as we will see. Note that the calculation of figure 2.1 assumes that Earth is perfectly spherical. Of course, it is not: due to rotation it bulges 31 kilometers (20 miles) at the equator.

2. Clouds are condensed water vapor, and thus are either water droplets or small ice crystals.

3. The current ice age began 2.58 million years ago and is known as the Quaternary glaciation. It is characterized by periods of increased ice coverage punctuated

by interglacial periods, such as the one we now live in, which has lasted for some 11,000 years. Since the last glacial maximum, sea levels have risen 120 meters (390 feet). We know that we are experiencing an interglacial period, rather than being outside an ice age, because the poles are capped with ice.

4. Albedo data for this section comes from A. K. Betts and J. H. Ball, "Albedo over the Boreal Forest," *Journal of Geophysical Research* 102 (1997): 28901–28910; P. R. Goode et al., "Earthshine Observations of the Earth's Reflectance," *Geophysical Research Letters* 28 (2001): 1671–1674; D. Hillel, *Environmental Soil Physics: Fundamentals, Applications, and Environmental Considerations* (London: Academic Press, 1998); M. Iqbal, *An Introduction to Solar Radiation* (New York: Academic Press, 1983); and E. Dobos, "Albedo," in *Encyclopedia of Soil Science*, 2nd ed., ed. R. Lal (Boca Raton, Fla.: CRC Press, 2006), 64–65. Ocean area and volume data are from B. W. Eakins and G. F. Sharman, "Volumes of the World's Oceans from ETOPO1," NOAA National Geographic Data Center, Boulder, Colo., 2010; and solar data are from National Aeronautics and Space Administration, "Sun: By the Numbers," Solar System Exploration, http://solarsystem.nasa.gov/planets/sun/facts.

5. Isaac Newton was well aware of the complications that arise due to the presence of a third body in a solar system. From his time to the present day, many astronomers have performed many calculations to determine the consequences of the extra body. For the real solar system, with dozens of gravitating bodies, accurate predictions require much computer number-crunching.

6. The reason is Kepler's second law of planetary motion, which tells us that an orbiting body moves faster when closer to the sun. (Think of a comet whipping past the sun when close, and slowing down when farther out.) So, because the perihelion occurs on January 3, Earth is moving fastest during the northern winter, and thus winter is shorter than summer. This is a good point to note the effect of axial tilt on solar power density: polar regions receive 42% of the power that equatorial regions get, averaged over a year. The poles would receive less power if there were no axial tilt.

7. World War I was hard on rising and eminent physicists. The young Englishman Henry Mosely, who carried out pioneering work in spectroscopy before the war, was killed at Gallipoli in the same year as Milankovitch began his famous work. A year later, in 1916, Karl Schwartzschild died of disease on the Russian front. This German physicist had found a solution to Einstein's new General Theory of Relativity just before his death.

8. For more on Milankovitch theory and its observation in the climatology records, see J. D. Hays, J. Imbrie, and N. J. Shackleton, "Variations in the Earth's Orbit: Pacemaker of the Ice Age," *Science* 194 (1976): 1121–1132; E. A. Kasatkina, O. I. Shumilov, and M. Krapiec, "On Periodicities in Long Term Climate Variations Near 68°N, 30°E," *Advances in Geoscience* 13 (2007): 25–29; J.

Laskar et al., "La2010: A New Orbital Solution for the Long-Term Motion of the Earth," *Astronomy and Astrophysics* 532 (2011): A89; and S. E. Sondergard, *Climate Balance: A Balanced and Realistic View of Climate Change* (Mustang, Okla.: Tate, 2009). The NASA websites on Milankovitch are also very educational: Steve Graham, "Milutin Milankovitch (1879–1958)," March 24, 2000, Earth Observatory, www.earthobservatory.nasa.gov/Features/Milankovitch; Holli Riebeek, "Paleoclimatology: Explaining the Evidence," May 9, 2006, Earth Observatory, www.earthobservatory.nasa.gov/Features/Paleoclimatology_Evidence. See also National Oceanic and Atmospheric Administration, "Astronomical Theory of Climate Change," National Climatic Data Center, www.ncdc.noaa.gov/paleo/milankovitch.html.

9. For more details on the physics of tides, see M. Denny, *The Science of Navigation: From Dead Reckoning to GPS* (Baltimore: Johns Hopkins University Press, 2012), chap. 1.

10. The water cycle is discussed in detail in J. R. Gat, *Isotope Hydrology: A Study of the Water Cycle* (London: Imperial College Press, 2010). For a less technical account, see U.S. Geological Survey, "The Water Cycle," USGS Water Science School, http://water.usgs.gov/edu/watercycle.html. Another useful introduction is "Water Cycle," Wikipedia, https://en.wikipedia.org/wiki/Water_cycle.

11. Earlier books of mine on the subjects of navigation and of bird migration have also needed an introductory chapter on aspects of the physical Earth. The emphasis, however, has been different in each book. Thus historical navigation has made use of tides, while migrating birds use the geomagnetic field to find their way; neither of these aspects is significant here. Interestingly, the one phenomenon that it has been necessary to describe in all three cases is the Coriolis effect.

12. There is no horizontal Coriolis force at the equator; there is a small vertical Coriolis force that can usually be ignored.

13. There are many readable references for ocean currents and circulations. See, for example, W. Broecker, *The Great Ocean Conveyor: Discovering the Trigger for Abrupt Climate Change* (Princeton, N.J.: Princeton University Press, 2010); R. A. Kerr, "A Slowing Cog in the North Atlantic Ocean's Climate Machine," *Science* 304 (2004): 371–372; S. Rahmstorf, "Rapid Climate Transitions in a Coupled Ocean–Atmosphere Model," *Nature*, November 3, 1994, 82–85; and K. A. Sverdrup, A. C. Duxbury, and A. B. Duxbury, *Fundamentals of Oceanography*, 5th ed. (New York: McGraw-Hill, 2006). For a nontechnical account, see National Oceanic and Atmospheric Administration, "Currents," http://oceanservice.noaa.gov/education/tutorial_currents/welcome.html.

14. El Niño translates from Spanish as "male child" or "Christ child" because of its December appearance off South America. La Niña translates as "female child."

15. Increased temperature causes the waters of the tropical Pacific to expand, and it has been observed that sea surface levels are higher during El Niño years by 0.5 meter (1.5 feet) or so.

16. The El Niño phenomenon is of broad interest, unsurprisingly, and so there is much reporting on the subject. For popular accounts, see J. D. Cox, *Weather for Dummies* (Indianapolis: Wiley, 2000), 132–137; National Oceanic and Atmospheric Administration, "El Niño," Science with NOAA Research, http://www.oar.noaa.gov/k12/html/elnino2.html; National Aeronautics and Space Administration, "El Niño," NASA Science/Earth, http://science.nasa.gov/earth-science/oceanography/ocean-earth-system/el-nino/; and "El Niño," Wikipedia, https://en.wikipedia.org/wiki/El_Ni%C3%B1o. For more technical accounts, see, for example, K. E. Trenberth and T. J. Hoar, "The 1990–1995 El Niño–Southern Oscillation Event: Longest on Record," *Geophysical Research Letters* 23 (1996): 57–60; and K. E. Trenberth et al., "Observations: Surface and Atmospheric Climate Change," in *Climate Change 2007: The Physical Science Basis; Contribution of Working Group 1 to the Fourth Assessment Report of the Intergovernmental Panel on Climate Change*, ed. S. Solomon et al. (Cambridge: Cambridge University Press, 2007), sec. 3.6.2. For a summary of other synoptic (large-scale) meteorological oscillations, see H. M. Mogil, *Extreme Weather* (New York: Black Dog & and Leventhal, 2007), chap. 19.

3. The Air We Breathe

1. There are a number of reasons why temperature and density are different functions of altitude in the troposphere. Thus the troposphere is heated from above by shortwave solar radiation and from below by longwave radiation. These are absorbed differently by the atmospheric constituents and penetrate differently because of the difference in air density at the surface and tropopause.

2. The atmosphere's main constituent, nitrogen gas, also absorbs UV radiation. The absorption spectra of nitrogen and ozone are different, however, and ozone absorbs some frequencies that nitrogen does not. UV radiation reaching the surface can be damaging to health; this is why ozone is important even though it exists only in trace amounts. Chlorofluorocarbons (CFCs), man-made chemicals, have been shown to deplete atmospheric ozone, and so their use has been heavily regulated since the late 1980s. The current hole in the ozone layer over Antarctica is attributed to CFCs, though the extent and degree of depletion are not well understood. The ozone hole is the subject of much current research and detailed observations.

3. We will see later that some thunderstorms can extend up into the lower stratosphere for a brief period of time. Also, jet streams can mix up tropospheric and stratospheric air.

4. We might have developed eyes that are sensitive to radio waves, which you can see from figure 3.2 suffer even lower attenuation through the atmosphere than do visible wavelengths, but the result would be very inferior vision. The intensity of radio waves from the sun is much less than the intensity of visible light, and the ability for lenses to resolve images relies on the lens being larger than the wavelength.

5. There are plenty of online sources for atmospheric composition. Much of my interpretation comes from Thomas W. Schlatter, "Atmospheric Composition and Vertical Structure," July 23, 2009, http://ruc.noaa.gov/AMB_Publications_bj/2009%20Schlatter_Atmospheric%20Composition%20and%20Vertical%20Structure_eae319MS-1.pdf; and Wikipedia, "Atmosphere of Earth," https://en.wikipedia.org/wiki/Atmosphere_of_Earth. The odd status of argon (how many people know that argon is the third most common gas?) and its origins are discussed in Rensselaeur Polytechnic Institute, "Argon Conclusion: Researchers Reassess Theories on Formation of Earth's Atmosphere," *ScienceDaily*, September 24, 2007, www.sciencedaily.com/releases/2007/09/070919131757.htm.

6. There are some serious exceptions to this observation about air currents not being much influenced by land. Thus we have already noted the strong and persistent winds in southern latitudes (the Roaring Forties, Furious Fifties, and Screaming Sixties) due to lack of land obstacles; these winds suggest that trade winds at other latitudes might be stronger if there were no land to impede them. Also, in the recent literature, there has appeared an alternative to the currently favored theory on the origin of the mild winters in northwestern Europe (that is, that they are a consequence of the Gulf Stream): westerly winds bouncing off the Rocky Mountains create warm sou'westers that account for most of the observed warming during the winter. See R. Seager et al., "Is the Gulf Stream Responsible for Europe's Mild Winters?" *Quarterly Journal of the Royal Meteorological Society* 128 (2002): 2563–2586; and R. Seager, "The Source of Europe's Mild Climate," *American Scientist* 94 (2006): 334–341.

7. Not to be confused with hurricanes—called cyclones in some parts of the world, as we will see. A cyclone is any counterclockwise-spiraling packet of air in the Northern Hemisphere (clockwise-spiraling south of the equator). An anticyclone is any packet of air that spirals in the opposite sense. It may seem confusing to recall that the Coriolis force leads to oceanic gyres that rotate *clockwise* in the Northern Hemisphere, opposite to cyclone rotation. In fact, these behaviors are consistent. Gyres are formed by easterly and westerly winds, at different latitudes, that are deflected by the Coriolis force. Cyclones are formed by winds heading toward a low-pressure center, which are deflected by the same force.

8. The ITCZ is not exactly at the equator; it is a ragged line that circles the globe within a few degrees of it, moving with the seasons (referring to northern seasons, a little north in the summer and south in the winter).

9. It may not always have been so in earlier epochs. It has been argued in the technical literature that during the Eocene and Cretaceous periods, the Hadley cells may have extended much farther north. There was no polar ice during these warm periods, so the temperature gradient from equator to poles was much shallower than at present and the force driving warm air poleward was reduced. Thus wind speed and so Coriolis deflection were much less. This argument is not nailed down beyond dispute—it is a hypothesis, not a tested theory—but I thought it worth airing (so to say). More details can be found, for example, at Harvard University, "Hadley Cells," http://www.seas.harvard.edu/climate/eli/research/equable/hadley.html.

10. Several folk myths have been offered to explain the odd name. One of the less plausible but more suggestive is that in the days of sailing ships, when Spaniards were transporting horses to their New World colonies, they became becalmed and ran out of water. To save water they upended their unfortunate equines into the briny deep.

11. Atmospheric circulation is a subject that, at its core, is very mathematical. Some accounts substitute math for lucidity. For readable explanations of varying depth, see J. D. Cox, *Weather for Dummies* (Indianapolis: Wiley, 2000), chap. 4; M. Pidwirny, *Understanding Physical Geography*, 3 parts (Kelowna, B.C.: Our Planet Earth, 2014), chap. 7; T. Schneider, "The General Circulation of the Atmosphere," *Annual Review of Earth and Planetary Science* 34 (2006): 655–688; and A. H. Strahler and A. Strahler, *Physical Geography: Science and Systems of the Human Environment*, 2nd ed. (New York: Wiley, 2003), chap. 7. Useful online explanations include Yochanan Kushnir, "General Circulation and Climate Zones," The Climate System, http://eesc.columbia.edu/courses/ees/climate/lectures/gen_circ/; National Weather Service, National Oceanic and Atmospheric Administration, "Global Circulations," www.srh.noaa.gov/jetstream/global/circ.html; and Met Office, "Global Circulation Patterns," January 7, 2016, www.metoffice.gov.uk/learning/learn-about-the-weather/how-weather-works/global-circulation-patterns. See also Y. Hu and Q. Fu, "Observed Poleward Expansion of the Hadley Circulation Since 1979," *Atmospheric Chemistry and Physics* 7 (2007): 5229–5236; and J. Huang and M. B. McElroy, "Contributions of the Hadley and Ferrel Circulations to the Energetics of the Atmosphere over the Past 32 Years," *Journal of Climate* 27 (2014): 2656–2666.

12. Some of the other jet streams travel westward (such as the equatorial jet, which sometimes arises at low latitudes); another is low-level, just a few hundred meters above the Pacific Ocean (Cox, *Weather for Dummies*); another is unique in that it has little to do with our surface weather, but occurs way up in the tenuous thermosphere, at 100-kilometer (62-mile) altitude (the Kármán line, you may recall), and it blows at very high speeds, around 400 kilometers (250 miles) per hour. For this latter jet, which was discovered only in the 1960s, see National

Aeronautics and Space Administration, "Earth's Two Jet Streams," http://www.nasa.gov/mission_pages/sunearth/news/gallery/atrex-jetstream-locations.html.

13. These jet streams are exploited by airlines. When flying east, the tailwind provided by a jet stream reduces airplane fuel consumption and journey time. The time difference flying across the Atlantic is about an hour.

14. Weather is characterized by phenomena that cover a wide range of length and timescales. These are not independent: smaller phenomena are usually also briefer. Thus microscale phenomena, such as turbulence and wind gusts, occur over a few meters or a few hundred meters and last for seconds or minutes. Mesoscale phenomena, such as thunderstorms, cover tens of kilometers with a duration of minutes to hours. Synoptic-scale phenomena such as cyclones cover thousands of kilometers and last for days or weeks. The largest phenomena such as trade winds are of planetary scale and change over weeks or months.

15. A number of useful websites provide nontechnical information about jet streams. See, for example, NASA, "Earth's Two Jet Streams"; and National Weather Service, National Oceanic and Atmospheric Administration, "The Jet Stream," www.srh.noaa.gov/jetstream/global/jet.html. For more detailed and technical accounts, see, for example, A. Gettelman, M. L. Salby, and F. Sassi, "Distribution and Influence of Convection in the Tropical Tropopause Region," *Journal of Geophysical Research* 107 (2002): 4080; E. Reiter, "Tropospheric Circulation and Jet Streams," in *Climate of the Free Atmosphere*, vol. 4 of *World Survey of Climatology*, ed. D. F. Rex (New York: Elsevier, 1969), chap. 4; and G. Zängl and K. P. Hoinka, "The Tropopause in the Polar Regions," *Journal of Climate* 14 (2001): 3117–3139.

4. Dynamic Planet

1. That timescale influences the nature of the problem we are investigating becomes apparent when we consider the rotation of Earth. This rotation introduces temperature oscillations with a period of 24 hours over all points of Earth. If we work with data that are averaged over 24 hours, then we cannot make predictions about weather phenomena that depend on this oscillation, such as sea breezes. If we work with data that are averaged over a year, then we cannot make predictions about seasonal phenomena.

2. The greenhouse analogy is chosen for its simplicity of calculation. The price paid is some misleading conclusions (if we are not careful) arising from the difference between glass and air. Thus glass is nearly a perfect absorber of longwave radiation, whereas air is not. Air can transfer heat by convection as well as by radiation. Atmospheric density (and hence absorption and other thermodynamic properties) varies with altitude. The basic result for both greenhouse and atmosphere—increasing surface temperature due to absorption of its blackbody

radiation aloft—is why we dub this phenomenon the "greenhouse effect." It is a useful shorthand, but ultimately the analogy is an oversimplification.

3. One of the concerns about rising atmospheric carbon dioxide levels is that it partly closes the atmospheric window shown in figures 3.2 and 4.2.

4. In the literature, it is conventional to refer to this type of calculation as "energy budget," rather than "power budget," hence the title of this section. I prefer "power budget" simply because the incoming solar power is constant in these models, but the distinction is minor. Power is energy per unit time, and so, for example, the power absorbed by the atmosphere in one second is a quantity of energy. Take your pick.

5. We saw a hint that this two-layer model of the atmosphere might be significant in the cloche model with two layers of glass—their equilibrium temperatures were different, with the glass nearer the surface being warmer.

6. For technical and popular accounts of the snowball Earth theory and observations, see M. Budyko, "The Effect of Solar Radiation Variations on the Climate of the Earth," *Tellus A* 21 (1969): 611–619; C. Dell'Amore, "'Snowball Earth' Confirmed: Ice Covered Equator" [online], *National Geographic*, March 4, 2010. http://news.nationalgeographic.com/news/2010/03/100304-snowball-earth-ice-global-warming/; and Encyclopaedia Britannica, s.v. "Snowball Earth hypothesis," http://www.britannica.com/science/Snowball-Earth-hypothesis. The faint young sun paradox and its possible resolutions are discussed in C. Karoff and H. Svensmark, "How Did the Sun Affect the Climate When Life Evolved on the Earth?" arXiv (2010): 1003.6043; D. Netburn, "Mystery of the 'Faint Young Sun Paradox' May Be Solved," *Los Angeles Times*, June 10, 2013; S. Reardon, "Titan Holds Clue to Faint Young Sun Paradox," *New Scientist*, January 2013; K. H. Schatten and A. S. Endal, "The Faint Young Sun-Climate Paradox: Volcanic Influences," *Geophysical Research Letters* 9 (1982): 1309–1311; and R. Wordsworth and R. Pierrehumbert, "Hydrogen-Nitrogen Greenhouse Warming in Earth's Early Atmosphere," *Science* 339 (2013): 64–67.

7. L. B. Larsen et al., "New Ice Core Evidence for a Volcanic Cause of the A.D. 536 Dust Veil," *Geophysical Research Letters* 35 (2008): L04708; H. M. Mogil, *Extreme Weather* (New York: Black Dog & Leventhal, 2007), 230, chap. 18; A. Witze and J. Kanipe, *Island on Fire: The Extraordinary Story of Laki, the Volcano That Turned Eighteenth-Century Europe Dark* (London: Profile Books, 2014); and G. D. Wood, *Tambora: The Eruption That Changed the World* (Princeton, N.J.: Princeton University Press, 2014).

8. For readable accounts of the climate effects of volcanic gas emissions, see J. LaPan, "Particles in Upper Atmosphere Slow Down Global Warming," July 25, 2011, National Aeronautics and Space Administration, http://www.nasa.gov/topics/earth/features/stratospheric-aerosols.html; and Volcanic Ashfall Impacts Working Group, "Ashfall Is the Most Widespread and Frequent Volcanic Hazard,"

February 2, 2015, U.S. Geological Survey, http://volcanoes.usgs.gov/vhp/tephra.html.

9. Environmentalists and some climatologists worry about the amount of fuel exhaust that airplanes deposit in the stratosphere because of the long time that such exhaust gases and particles stays up there.

10. According to K. C. Harper, "Hoping to catch a thunderstorm in a large-grid model is like trying to catch small fish in a net with large holes" ("Weather Forecasting by Numerical Methods," in *Discoveries in Modern Science: Exploration, Invention, Technology*, ed. J. Trefil [Farmington Mills, Mich.: Macmillan, 2015]). Even today, individual thunderstorms are not well resolved in NWP models.

11. The Met Office, the national weather service of the United Kingdom, has a large climate model computing facility that is matched by perhaps only a dozen in other parts of the world: as of 2014, it consists of two CRAY 3TE supercomputers, each with 900 processors and multiterabyte data storage. Of course, these numbers will be out of date within a few years.

12. For more detailed or technical accounts of GCMs, see, for example, G. Flato et al., "Evaluation of Climate Models," in *Climate Change 2013: The Physical Science Basis; Contribution of Working Group 1 to the Fifth Assessment Report of the Intergovernmental Panel on Climate Change*, ed. T. F. Stocker et al. (Cambridge: Cambridge University Press, 2013), 741–865; K. McGuffie and A. Henderson-Sellers, *A Climate Model Primer* (New York: Wiley, 2013); B. J. Soden and I. M. Held, "An Assessment of Climate Feedbacks in Coupled Ocean–Atmosphere Models," *Journal of Climate* 19 (2006) 3354–3360; and S. Weart, "General Circulation Models of the Atmosphere," in *Discoveries in Modern Science*, ed. Trefil, 41–46. A useful introduction is "General Circulation Model," Wikipedia, https://en.wikipedia.org/wiki/General_Circulation_Model. By the year 2000, there were a dozen major teams with climate models around the world. Each model requires more than a million lines of computer code.

13. This statement is not quite true. There are specialized climate models designed specifically to predict the evolution of the ozone hole, which necessitates modeling the stratosphere.

14. A "sliding window" for time-series data is one that moves in time. Thus if the climate for the year 2000 is estimated by averaging the weather data from 1985 to 2015, then the climate for 2001 is found by averaging data from 1986 to 2016, and so on. In this case the weather data are being smoothed by applying a 31-year sliding window.

15. Other proxy data sources include pollen, insects, stalactites and stalagmites, and historical records of harvest dates and harbor ice-free dates. For a description of these proxies, and for the data itself, see National Oceanic and Atmospheric Administration, National Centers for Environmental Information, "Paleoclimatology Datasets," www.ncdc.noaa.gov/data-access/paleoclimatology-data/datasets.

16. There is another strong argument for a human origin of global warming: carbon isotope ratios fingerprint human rather than natural causes for atmospheric carbon dioxide levels.

17. Data for this section come from D. J. Baker, "Climate Change," in *Discoveries in Modern Science*, ed. Trefil, 188–189; S. P. Huang, H. N. Pollack, and P.-Y. Shen, "A Late Quaternary Climate Reconstruction Based on Borehole Heat Flux Data, Borehole Temperature Data, and the Instrumental Record," *Geophysical Research Letters* 35 (2008): L13703; P. W. Leclercq and J. Oerlemans, "Global and Hemispheric Temperature Reconstruction from Glacier Length Fluctuations," *Climate Dynamics* 38 (2012): 1065–1079; and S. Solomon et al., eds., *Climate Change 2007: The Scientific Basis; Contribution of Working Group 1 Contribution to the Fourth Assessment Report of the Intergovernmental Panel on Climate Change* (Cambridge: Cambridge University Press, 2007). The physics of our atmosphere is largely but not completely understood. In particular, climate model accuracy is limited by our incomplete understanding of how atmospheric processes vary with altitude. This area, and others that impinge on climate model prediction accuracy, are being actively researched. See, for example, M. Caldwell, "Unravelling Our Atmosphere," *Physics World* 27 (2014): 36–40.

18. A few skeptics remain about the shape of Earth—they think it's flat. See http://theflatEarthsociety.org/cms/. But not all skeptics are of this ilk. Questioning theories is part of being a scientist and interpreting statistics is not always straightforward. Thus Mogil, while not doubting the reality of global warming and of the human contribution to it, urges caution in interpreting the data when predicting the consequences, in *Extreme Weather*. Thus, for example, the damage caused by thunderstorms is increasing over recent decades, but careful analysis shows that this is largely due to a greater population at risk and not to an increase in thunderstorm severity. More on the pitfalls of statistical interpretation are in chapter 6.

19. "Heart of the Matter" [editorial], *Nature*, July 28, 2011, 424.

20. For the heating of deep-sea waters, see J.-P. Gattuso et al. "Contrasting Futures for Ocean and Society from Different Anthropogenic CO_2 Emission Scenarios," *Science* 349 (2015): 45; V. Guemas et al., "Retrospective Prediction of the Global Warming Slowdown in the Past Decade," *Nature Climate Change* 3 (2013): 649–653; and S. Levitus et al., "World Ocean Heat Content and Thermosteric Sea Level Change (0–2000 m), 1955–2010," *Geophysical Research Letters* 39 (2012): L10603.

21. Thus increased temperature causes permafrost to melt, which releases trapped methane into the atmosphere, which further increases temperature, and so on.

22. Readable accounts of future climate changes and their likely consequences are those of Baker, "Climate Change," 188–189; and A. Weaver, *Generation Us: The Challenge of Global Warming* (Victoria, B.C.: Orca, 2011).

5. Oceans of Data

1. Experienced human observers are historically superior to automated measurement systems when it comes to providing a horizon-to-horizon estimation of current weather conditions, though the automated systems are getting better. Also they are more reliable and do not tire.

2. Radar was instrumental (pardon the pun) in winning the Battle of Britain and later during the bombing of German cities.

3. That is, a radar will take about five minutes to scan its search volume completely. The radar beam begins by making one complete rotation scanning the sky near the horizon, then the beam is elevated by one beam width for the second rotation, then elevated again for the third rotation, and so on, until the required search volume is covered. Then the search repeats.

4. For a detailed history of the evolution and development of weather radar, see R. Whiton et al., "History of Operational Use of Weather Radar by U.S. Weather Services, Part II: Development of Operational Doppler Weather Radars," *Weather and Forecasting* 13 (1988): 244–252. For weather radar, see M. Denny, *Blip, Ping, and Buzz: Making Sense of Radar and Sonar* (Baltimore: Johns Hopkins University Press, 2007), 157–161; and H. M. Mogil, *Extreme Weather* (New York: Black Dog & Leventhal, 2007), 90–91.

5. Readers who are physicists may be intrigued to learn that the ascent rate is so constant given that the air becomes rarefied and the balloon radius expands significantly as it rises. I also was intrigued by this and wrote a small paper explaining why it occurs. See M. Denny, "Weather Balloon Ascent Rate," *Physics Teacher* 54 (2016): 268–271.

6. For more information on radiosondes and weather balloons, see I. Durre, R. S. Vose, and D. B. Wuertz, "Overview of the Integrated Global Radiosonde Archive," *Journal of Climate* 19 (2006): 53–68; and National Oceanic and Atmospheric Administration, National Weather Service, National Oceanic and Atmospheric Administration, "Radiosonde Observations," http://www.ua.nws.noaa.gov/factsheet.htm.

7. There is a western European equivalent to POES: the three MetOp weather satellites that are in polar orbit. Indeed, there is much cooperation between Europe and the United States in the polar weather satellite programs. Many other countries—Canada, Japan, and Russia among them—employ such weather satellites.

8. Often, the *interpretation* of the data may vary from country to country or from one interest group to the next, based on factors other than pure science, but this does not interfere with clear need for cooperation in obtaining the data.

9. China Meteorological Administration, "International Exchange and Cooperation," National Marine Information and Data Service, http://www.nmdis.gov.cn/english/gjhz/gjhz.html.

10. American Meteorological Society, "Full and Open Access to Data," December 4, 2013, https://www.ametsoc.org/ams/index.cfm/about-ams/ams-statements/statements-of-the-ams-in-force/full-and-open-access-to-data/.

11. Despite standardization, there are differences in the quality and quantity of data contributed from different sources. These data need to be treated (homogenized) before they can be input to numerical weather prediction computer programs.

6. Statistically Speaking

1. Ahem, not politically.

2. For handedness statistics, see M. Papadatou-Pastou et al., "Sex Differences in Left-Handedness: A Meta-analysis of 144 Studies," *Psychological Bulletin* 134 (2008): 677–699.

3. Unrelated except in the type of randomness. The binomial distribution applies to sequences of independent two-outcome experiments with fixed probability for each outcome.

4. For the Prussian cavalry example, see D. F. Andrews and A. M. Herzberg, *Data: A Collection of Problems from Many Fields for the Student and Research Worker* (New York: Springer, 1985).

5. You may recall the caption to figure 4.8, which included 95%-confidence error bars. We see now that this confidence level corresponds to two standard deviations.

6. Data for height distribution is from U.S. Census Bureau, *Statistical Abstract of the United States: 2012* (Washington, D.C.: Government Printing Office, 2012), https://www.census.gov/library/publications/2011/compendia/statab/131ed.html.

7. Those readers who are quantum physicists will know that, where quantum effects arise, they contribute to statistical measurements even in the absence of measurement error. The point I am making here is that statistical uncertainty, whatever its cause, can be quantified.

8. Statistically speaking, most readers of this book will be American and so, I feel sure you will have noticed, here I employ the archaic Fahrenheit temperature scale in addition to (added parenthetically after) the more modern and scientific Celsius or centigrade scale adopted everywhere else in the world.

9. For the mathematically inclined, I note here that equation (6.1) has two attractors, both unstable, at $x = 1.96628783$ and at $x = -0.96628783$.

10. For the original, typewritten, paper, see Edward Lorenz, "Predictability: Does the Flap of a Butterfly's Wings in Brazil Set Off a Tornado in Texas?" (paper presented at the 139th meeting of the American Association for the Advancement of Science, December 29, 1972, Boston), http://eaps4.mit.edu/research

/Lorenz/Butterfly_1972.pdf. Lorenz is quoted in C. M. Danforth, "Chaos in an Atmosphere Hanging on a Wall," Mathematics of Planet Earth, http://mpe2013.org/2013/03/17/chaos-in-an-atmosphere-hanging-on-a-wall/.

11. In fact, sample size is not the whole story in this case. The temperature for Washington State involves large-scale, slow-moving weather systems whereas those for a smaller area such as Seattle are inherently less sluggish, so predicting temperatures over a large area is easier—less error prone—than predicting local temperatures. But the point remains: increasing sample size reduces measurement errors.

7. A Condensed Account of Clouds, Rain, and Snow

1. Cloud classification is outlined in the section "Clouding the Issue."

2. Salt particles originate from bubbles that burst on the sea surface; the resulting tiny airborne droplets then lose their water through evaporation. Pollutants are cleaned from the atmosphere by precipitation: pollutant particles act as nuclei for water vapor condensation and so when the droplets grow into raindrops they return the pollutant particle to the surface.

3. Icing of airplanes is heaviest when the air temperature is near –10°C (14°F).

4. When the temperature is –40° it is unnecessary to specify centigrade or Fahrenheit, because they are the same: –40°C = –40°F. Only when the temperature is below –15°C (5°F) does ice formation exceed water droplet formation in clouds.

5. A lower density of condensation nuclei over oceans means that maritime cloud droplets are fewer and larger than land cloud droplets. They grow faster (due to higher average water vapor density at low altitudes). In addition, the presence of ice in a cloud enhances (or is the sole cause of) precipitation, and ice crystals occur at higher temperatures in clouds with precipitation-size drops than in clouds that do not contain these. Maritime regions are so depleted of condensation nuclei that ship tracks can be detected by observing the cloud trail resulting from their emissions.

6. The information in this section is drawn mostly from B. J. Mason, "The Oceans as a Source of Cloud-Forming Nuclei," *Pure and Applied Geophysics* 36 (1957): 148–155; H. M. Mogil, *Extreme Weather* (New York: Black Dog & Leventhal, 2007), 214; T. Nishikawa, S. Maruyama, and S. Sakai, "Radiative Heat Transfer Analysis Within Three-Dimensional Clouds Subjected to Solar and Sky Irradiation," *Journal of the Atmospheric Sciences* 61 (2004): 3125–3133; A. Rangno, "Classification of Clouds," in *Discoveries in Modern Science: Exploration, Invention, Technology*, ed. J. Trefil (Farmington Mills, Mich.: Macmillan, 2015), 199–201; and National Aeronautics and Space Administration, "Cloud Droplets and Rain

Drops," http://scool.larc.nasa.gov/lesson_plans/CloudDropletsRainDrops.pdf; "Clouds and Radiation," http://earthobservatory.nasa.gov/Features/Clouds/; and "Cloud Climatology: Global Distribution and Character of Clouds," http://www.giss.nasa.gov/research/briefs/rossow_01/distrib.html. I draw your attention also to the excellent National Weather Service, National Oceanic and Atmospheric Administration, "Clouds," http://www.srh.weather.gov/jetstream/clouds/clouds_intro.html, which is part of its site Jetstream—An Online School for Weather.

7. The French evolutionist Jean-Baptiste Lamarck had earlier published his own scheme for clouds but it made little impact—less than that of his widely discussed but wrong theory of evolution. For more about Luke Howard and how he came to develop his ideas about clouds, see R. Hamblyn, *The Invention of Clouds: How an Amateur Meteorologist Forged the Language of the Skies* (New York: Picador, 2001).

8. See, for example, National Weather Service, National Oceanic and Atmospheric Administration, "NWS Cloud Chart," www.srh.noaa.gov/jetstream/clouds/cloudchart.htm; and Cloud Appreciation Society, photo gallery, https://cloudappreciationsociety.org/gallery/. Also useful are Met Office, "Cloud Types and Pronunciation" (chart) and "Cloud-Spotting Guide" (video), http://www.metoffice.gov.uk/learning/clouds/cloud-spotting-guide.

9. Cloud droplet characteristics and distribution are discussed in, for example, L. F. Radke, J. A. Coakley, and M. D. King, "Direct and Remote Sensing Observations of the Effects of Ships on Clouds," *Science* 246 (1989): 1146–1149; and D. Rosenfeld and I. M. Lensky, "Satellite-Based Insights into Precipitation Formation Processes in Continental and Maritime Convective Clouds," *Bulletin of the American Meteorological Society* 79 (1998): 2457–2476.

10. There is a great deal of online information about fog and fog types. See, for example, fog classification in National Weather Service, National Oceanic and Atmospheric Administration, "Fog Resources," www.nws.noaa.gov/om/fog; Met Office, "What Is Fog?" http://www.metoffice.gov.uk/learning/fog; and "Fog," Wikipedia, https://en.wikipedia.org/wiki/Fog.

11. This average rate corresponds to approximately one raindrop per square meter per second. Well, I feel it is important for you to know.

12. The claims for driest and wettest places on Earth are disputed and depend in part on statistics. (Average rate or peak rate? Average peak rate or greatest recorded extreme? Average over one decade or since records began?) Another aspect of the dispute is connected with regional pride or tourist advertising. For a list of the 10 wettest places on Earth, see, for example, A. House, "The Top 10 Wettest Places on Earth," *Daily Telegraph*, August 18, 2014.

13. Rainfall rate and raindrop dynamics are discussed in B. F. Edwards, J. W. Wilder, and E. E. Scime, "Dynamics of Falling Raindrops," *European Journal of Physics* 22 (2001): 113–118; and J. A. Smith et al., "Variability of Rainfall Rate and

Raindrop Size Distributions in Heavy Rain," *Water Resources Research* 45 (2009): WO4430.

14. The largest hailstone recorded to date was the one that landed on Vivian, South Dakota, on July 23, 2010: it was 20 centimeters (8 inches) across and weighed nearly 1 kilogram (2 pounds). Hail is most common in the interior of continents at temperate latitudes. See "Hail," Wikipedia, https://en.wikipedia.org/wiki/Hail; New World Encyclopedia, s.v. "Hail," www.newworldencyclopedia.org/entry/Hail; National Weather Service, National Oceanic and Atmospheric Administration, "Thunderstorm Hazards—Hail," www.srh.noaa.gov/srh/jetstream/tstorms/hail.html; and N. J. Doesken, "Hail, Hail Hail! The Summertime Hazard of Eastern Colorado," *Colorado Climate* 17, no. 7 (1994), http://www.cocorahs.org/media/docs/hail_1994.pdf.

15. For a more detailed look at snowflake form and formation, R. Cowan explains how scientists can now model snowflake growth on a computer "after a flurry of attempts" ("Snowflake Growth Successfully Modeled from Physical Laws" [online], *Scientific American,* March 16, 2012, http://www.scientificamerican.com/article/how-do-snowflakes-form/). See also M. Peplow, "Snowflakes Made Easy," *Nature*, December 31, 2014. A more complete account of precipitation in general is that of P. K. Wang, *Physics and Dynamics of Clouds and Precipitation* (Cambridge: Cambridge University Press, 2013).

16. I have found no better popular explanation of thunderstorms than that of National Oceanic and Atmospheric Administration, National Severe Storms Laboratory, "Severe Weather 101: Thunderstorm Basics, http://www.nssl.noaa.gov/education/svrwx101/thunderstorms/. It is worth emphasizing that there are two types of wind shear: that arising from different wind directions and that arising from different wind speeds. It is the latter type that is primarily responsible for the separation of updraft and downdraft regions in thunderstorms.

17. Above the thundercloud, *upper-atmosphere lightning* is occasionally seen arcing across the stratosphere or ionosphere. This lightning is distinct from thunderstorm lightning, which is confined to the troposphere and generated by collisions between ice particles.

18. For data on lightning see M. Akita et al., "Effects of Charge Distribution in Thunderstorms on Lightning Propagation Paths in Darwin, Australia," *Journal of the Atmospheric Sciences* 68 (2011): 719–726; M. Denny, *Lights On! The Science of Power Generation* (Baltimore: Johns Hopkins University Press, 2013); Encyclopaedia Britannica (1998), s.v. "Electrical Charge Distribution in a Thunderstorm"; T. C. Marshall and M. Stolzenburg, "Voltages Inside and Just Above Thunderstorms," *Journal of Geophysical Research* 106 (2001): 4745–4768; National Oceanic and Atmospheric Administration, National Severe Storms Laboratory, "Severe Weather 101: Lightning Basics," http://www.nssl.noaa.gov/education/svrwx101/lightning/; and "Lightning," Wikipedia, https://en.wikipedia.org/wiki/Lightning.

19. R. G. Roble and I. Tzur, "The Global Atmospheric-Electric Circuit," in *The Earth's Electrical Environment*, ed. National Research Council (Washington, D.C.: National Academies Press, 1986), 206–231.

8. Weather Mechanisms

1. Of course, I am skipping over many of the complications by restricting attention to simple *physical* laws and sweeping under the carpet many complicated *biological* phenomena. Thus, for example, a tree is (to a meteorologist) a very efficient machine for turning groundwater into atmospheric water vapor; it is (to a climatologist) a key constituent of the carbon cycle, taking carbon dioxide out of the atmosphere and putting it into the ground, as coal. I will ask you to simply accept the assertions that water vapor and carbon dioxide exist in the atmosphere and to not ask for details of how they got there or how they are removed.

2. The First Law is simply a statement of energy conservation—that the energy in a thermodynamic system (such as a parcel of air) can change form but not amount. In particular, it states that the heat energy that is added to the system equals the work done by the system plus the change in internal energy of the system. Thus, for example, heating a parcel of air will cause its volume to increase (which does work by displacing other air) and its temperature to change.

3. It is conventional in atmospheric-physics teaching to refer to a big bunch of air as a *parcel* or *packet*. The quantity is unspecified because in reality it is varied. In this case, I am referring to a volume of air near Earth's surface that rises as a result of being heated by the sun. A parcel of air is small enough so that it is sensible to talk about its temperature and pressure, whereas larger volumes of air have different temperatures and pressures at different locations.

4. There are several reasons why air rises apart from heating of the surface (convection). A horizontal wind is pushed upward when it encounters mountains; a warm front ascends cold surface air that it encounters; and a cold front pushes warm air upward.

5. Here I am omitting temporarily the complication that arises when the parcel is a different temperature from the surrounding air. Let us say that all the air is the same temperature. If the parcel of air were warmer than its surroundings, then it would rise due to buoyancy force rather than to pressure differences.

6. *Adiabatic* means "without heat transfer." Thus the calculation that determines dry lapse rate assumes that the parcel of air that is rising does not transfer heat to the surrounding air. Given that air is a poor conductor of heat (think of those voluminous winter jackets) and that the volume of an air parcel is usually enormous, this assumption is pretty good. Note that the dry lapse rate applies for dry air and for air that contains water vapor, as long as it is not saturated.

7. If the temperature is increasing with altitude, then it is lapsing (decreasing) at a negative rate.

8. Cool air below warm air is unusual because of the pressure profile of the atmosphere: air pressure falls more or less exponentially with increasing altitude. Because of this pressure fall-off, we expect thermodynamically that temperature will decrease with altitude (rising air will expand and cool).

9. When we consider only the thermodynamics of a parcel of air (ignoring its other forms of energy, such as gravitational potential energy), then the thermodynamic energy balance is described via the First Law of Thermodynamics.

10. For those readers who are interested in the details, the angular momentum magnitude is given by $kmr^2\omega$, where m is the total wheel mass, k is a constant that depends on the mass distribution of the wheel, r is wheel radius, and ω is wheel angular speed. The angular momentum direction is found by mentally curling the fingers of your right hand around the axle (in the same sense as the rotation, clockwise or counterclockwise) and extending your thumb—your thumb then points in the angular momentum direction.

11. I remember once seeing, in an Italian restaurant, waiters pouring bottles of wine into a display barrel. They spun the upturned bottles to induce a spin in the wine as it drained. They said the bottle emptied faster that way.

12. Much of the data for this section comes from *Encyclopaedia Britannica*, s.v. "Dust Devil." See also R. D. Lorenz and M. J. Myers, "Dust Devil Hazard to Aviation: A Review of United States Air Accident Reports," *Journal of Meteorology* 30 (2005): 178–184; P. C. Sinclair, "Some Preliminary Dust Devil Measurements," *Monthly Weather Review* 92 (1964): 363–367; and R. E. Wyett, "Pressure Drop in a Dust Devil," *Monthly Weather Review* 82 (1954): 7–8.

13. In fact, there are more tornadoes per unit area in the United Kingdom each year than there are in the United States, as noted in H. M. Mogil, *Extreme Weather* (New York: Black Dog & Leventhal, 2007), 105–106. This may be very surprising news to anyone in the United Kingdom until you realize that the British tornadoes are much, much weaker than American ones and cause far less damage.

14. The Fujita scale is associated with wind speeds of the tornado funnel at ground level. Thus an F1 tornado has wind speeds of around 90 miles per hour, whereas F5 speeds are 290 miles per hour (plus or minus; some exceed 300 miles per hour!). Above surface level, the speeds are higher than these figures. Now superseded by the Enhanced Fujita scale, denoted EF, which is less subjective and includes more aspects of tornado physics, the old scale was used for 30 years.

15. Thus Hurricane Andrew (1992) spawned damaging tornadoes in Louisiana and Mississippi, but only one in Florida. The much weaker Hurricane Danny (1985) gave rise to many powerful inland tornadoes.

16. It is difficult to obtain measurements of atmospheric pressure inside a tornado because measuring instruments tend to get destroyed by the tornado. The record pressure drop so far measured is 194 millibars.

17. This notion of how circulation tubes are generated has some photographic evidence to support it. The upside-down *U* (see figure 8.8) can be visible if condensation occurs, as shown in H. B. Bluestein, "More Observations of Small Funnel Clouds and Other Tubular Clouds," *Monthly Weather Review* 133 (2005): 3714–3720.

18. For a readable introduction to tornado physics, more accessible than many technical accounts, see D. E. Neuenschwander, "The Physics of Tornadoes," *SPS Observer*, Fall 2011, 2–17. For more detailed technical and historical articles, see, for example, C. A. Doswell III, "Historical Overview of Severe Convective Storms Research," *Electronic Journal of Severe Storms Meteorology* 2 (2007): 1–25; R. Edwards, "Tropical Cyclone Tornadoes: A Review of Knowledge in Research and Prediction," *Electronic Journal of Severe Storms Meteorology* 7 (2012): 1–11; and R. Edwards et al., "Storm Prediction Center Forecasting Issues Relating to the 3 May 1999 Tornado Outbreak," *American Meteorological Society* 17 (2002): 544–558. See also S. Perkins, "Tornado Alley, U.S.A," *Science News* 161 (2002): 296; and such informative websites as Storm Prediction Center, National Oceanic and Atmospheric Administration, "The Online Tornado FAQ," http://www.spc.noaa.gov/faq/tornado/, and National Severe Storms Laboratory, "Severe Weather 101: Tornado Basics," http://www.nssl.noaa.gov/education/svrwx101/tornadoes/. And then there are those cool tornado videos on YouTube.

19. The hurricane data is from National Oceanic and Atmospheric Administration, "Hurricanes," June 2015, http://www.education.noaa.gov/Weather_and_Atmosphere/Hurricanes.html, and "Hurricanes," July 12, 2004, http://www.oar.noaa.gov/k12/html/hurricanes2.html. The world electricity-generating capacity is from M. Denny, *Lights On! The Science of Power Generation* (Baltimore: Johns Hopkins University Press, 2013). If such a hurricane lasts a week, then it consumes 3.6×10^{20} joules of energy—that's 87,000 megatons TNT equivalent.

20. The Galveston hurricane (probably category 4) took place in 1900, before satellite technology provided us with a 24/7 eye on these superstorms—hence the high casualties.

21. In addition to damage caused directly by high winds and damage caused by storm surge, hurricanes can do a great deal of damage by generating flooding and landslides, as occurred in Central America with Hurricane Mitch in 1998.

22. This equality between pressure gradient force and centrifugal force is called *cyclostrophic balance* in meteorology textbooks. The eye is clear because of its sinking air, which inhibits cloud formation.

23. To a physicist, hurricanes are almost perfect heat engines, powered by the difference in temperature between the ocean and the air above it. There are many

popular-level accounts of hurricanes—unsurprisingly, given the effect they have when they make landfall. I particularly like University of Rhode Island, Graduate School of Oceanography, "Hurricanes: Science and Society" http://www.hurricanescience.org/science/science/. Also, for impressive images of hurricanes, see National Aeronautics and Space Administration, "Hurricanes and Tropical Storms," https://www.nasa.gov/mission_pages/hurricanes/main/index.html. For an accessible article on the physics of hurricanes, see R. Smith, "Hurricane Force," *Physics World*, June 2006, 32–37 and references.

9. Weather Extremes

1. A *black swan event* is an unpredicted outlier, a rare event that happens more often than we might suppose perhaps because of a misunderstanding of the underlying statistics. Thus many financial meltdowns, such as the banking crisis of 2008, are considered to be black swan events because we do not yet possess a full understanding of economic theory; weather catastrophes such as Hurricane Katrina might fall in the same category owing to incomplete understanding of the underlying mechanisms of global warming.

2. The information on heat waves past and future was gathered from nontechnical accounts for general readers, including G. Brücker, "Vulnerable Populations: Lessons Learnt from the Summer 2003 Heat Wave in Europe," *Eurosurveillance* 10 (2005); H. Hoag, "Russian Summer Tops 'Universal' Heatwave Index," *Nature*, October 29, 2014; E. Klinenberg, *Heat Wave: A Social Autopsy of Disaster in Chicago* (Chicago: University of Chicago Press, 2002); and O. Milman, "Heatwave Frequency 'Surpasses Levels Previously Predicted for 2030,'" *Guardian*, February 17, 2014, as well as from technical papers, including D. Coumou and A. Robinson, "Historic and Future Increase in the Global Land Area Affected by Monthly Heat Extremes," *Environmental Research Letters* 8 (2013): 034018; W. L. Kenney, D. H. Craighead, and L. M. Alexander, "Heat Waves, Aging, and Human Cardiovascular Health," *Medicine and Science in Sports and Exercise* 46 (2014): 1891–1899; S. E. Perkins and L. V. Alexander, "On the Measurement of Heat Waves," *Journal of Climate* 26 (2013): 4500–4517; and S. Russo, "Magnitude of Extreme Heat Waves in Present Climate and Their Projection in a Warming World," *Journal of Geophysical Research* 119 (2014): 12500–12512.

3. The total number of excess deaths during the European heat wave of 2003 was calculated at 70,000 by J.-M. Robine et al., "Death Toll Exceeded 70,000 in Europe During the Summer of 2003," *Comptes Rendus Biologies* 331 (2008): 171–178.

4. The dust bowl in the American Midwest was due in large part to the high frequency of heat waves that occurred there in the 1930s. Intense drought depleted

soil moisture and so reduced the moderating effects of evaporation—an example of positive feedback.

5. In Australia, the cities of Adelaide, Canberra, and Melbourne estimated that they would suffer a certain number of hot days over the period 2000 to 2030. In fact, they all "used up" all their anticipated hot days by 2009. It is important for local and national governments to know in advance what to expect (long term as well as in the immediate future) so they can plan mitigation, for example, with firefighting resources, air-conditioning installation and maintenance, and medical resources. An example of the far-reaching effects of a heat wave: the chronically ill and elderly often take prescription medication that interferes with their ability to dissipate heat, so during a heat wave they may have to be moved or grouped in air-conditioned facilities.

6. An amusing example of infrastructure for dealing with heat—or not, in this case—is that of Redford Barracks in Edinburgh, Scotland. These barracks were constructed in the nineteenth century at the same time as the British Army planned to build barracks for its soldiers in Calcutta, India. According to Edinburgh folklore, spectacular bureaucratic incompetence led to the plans for the two barracks getting mixed up. Whether this is true or not, it is certainly the case that the Scottish buildings are constructed to dissipate heat, with airy marble corridors, high ceilings, and large windows. Presumably, the heating bill in the frigid Scottish winters is enormous. I have been unable to establish whether or not the other barracks, in India, were ever constructed. If so, they would have been typically Scottish, with thick walls, heavily insulated roofs, and small windows, to conserve heat—in one of the hottest countries on Earth.

7. The exact time when peak oil occurs is a point in dispute, you may not be surprised to learn. It is likely to be within a decade or so, perhaps a little longer, or perhaps it has already occurred. Beyond peak oil, the supply dwindles, by definition. See M. Denny, *Lights On! The Science of Power Generation* (Baltimore: Johns Hopkins University Press, 2013), chap. 5.

8. H. M. Mogil, *Extreme Weather* (New York: Black Dog & Leventhal, 2007), 191. The water crisis in the Middle East is discussed in C. Arsenault, "Risk of Water Wars Rises with Scarcity," Al Jazeera, August 26, 2012; and the potential for future conflict over water resources in W. Barnaby, "Do Nations Go to War over Water?" *Nature*, March 19, 2009, 282–283; G. Dyer, *Climate Wars: The Fight for Survival as the World Overheats* (London: Oneworld, 2008); and S. Harris, "Water Wars," *Foreign Policy*, September 18, 2014. For more on peak water, see P. H. Gleick and M. Palaniappan, "Peak Water Limits to Freshwater Withdrawal and Use," *Proceedings of the National Academy of Sciences of the USA* 107 (2010): 11155–11162.

9. Not so much at the southern horse latitudes because there is much less land, and fewer people, in this region.

10. See, for example, P. Bowes, "California Drought: Will the Golden State Turn Brown?" BBC News, April 6, 2015; A. Holpuch, "Drought-Stricken California Only Has One Year of Water Left, NASA Scientist Warns," *Guardian*, March 16, 2015; and "Governor Brown Issues Order to Continue Water Savings as Drought Persists," May 9, 2016, https://www.gov.ca.gov/news.php?id=19408. The water shortage in the western United States may well spread beyond California in the coming decades, according to E. S. Povich, "Drought Is Not Just a California Problem," *USA Today*, April 19, 2015.

11. Thus, for example, the 10-year averages for the United States place cold snaps at the bottom of the table for weather fatalities among humans, at about 25 per year; heat waves are at the top, with five times the number. In between come floods, lightning, tornadoes, hurricanes, other winds, and rip currents. See National Weather Service, National Oceanic and Atmospheric Administration, "Weather Fatalities," www.nws.noaa.gov/om/hazstats.shtml. Note, though, that overall winter-related deaths may exceed those of the summer (due, for example, to extra traffic fatalities on icy roads, or a spike in fatal heart attacks while clearing snow). See D. Rice, "Killer Cold: Winter Is Deadlier than Summer in U.S.," *USA Today*, July 30, 2014.

12. For more on flooding in general, and on the truly awful Chinese floods of 1931 in particular, see C. D. Ahrens and P. Samson, *Extreme Weather and Climate* (Belmont, Calif.: Brooks/Cole, 2011); S. E. Mambretti, ed., *Flood Risk Assessment and Management* (Southampton, Eng.: WIT Press, 2012); C. Marshall, "Global Flood Toll to Triple by 2030," BBC News, March 4, 2015; A. N. Penna and J. S. Rivers, *Natural Disasters in a Global Environment* (Chichester, Eng.: Wiley-Blackwell, 2013); D. A. Pietz, *Engineering the State: The Huai River and Reconstruction in Nationalist China, 1927–1937* (New York: Routledge, 2002); and National Oceanic and Atmospheric Administration, "NOAA's Top Global Weather, Water, and Climate Events of the 20th Century," www.noaanews.noaa.gov/stories/s334b.htm.

13. In 1953, meteorologists began to name hurricanes, using female names (I do not want to know what the psychologists or sociologists would make of the gender bias); from 1979, both male and female names have been used.

14. Of the 875 hurricanes that were generated in the Atlantic Ocean between 1851 and 2010, only 33% reached land. The rest petered out or were reduced from hurricane status before they made it to a coast.

15. For more data than you would ever need about hurricanes that made landfall in the United States in general, and about Hurricane Katrina in particular, see E. S. Blake and E. J. Gibney, "The Deadliest, Costliest and Most Intense United States Tropical Cyclones from 1851 to 2010 (and Other Frequently Requested Hurricane Facts)," NOAA Technical Memorandum NWS NHC-6, National Weather Service, National Hurricane Center, Miami, Fla., 2011; and J. L. Beven et al., "Atlantic Hurricane Season of 2005," *Monthly Weather Review* 136 (2008): 1109–1173.

16. See, for example, J. Williams, "Doppler Radar Measures 318 mph Wind in Tornado," *USA Today*, May 17, 2005.

17. It is worth emphasizing that the losses here are financial, reflecting the development of the infrastructure of the countries affected. As always, the losses in human life are small in comparison with the losses in less developed countries to equivalent storms. Thus the Bhola cyclone in November 1970 killed between 300,000 and 500,000 people as winds of up to 115 mph powered a surge that flooded the Ganges Delta in East Bengal (now Bangladesh).

18. For more about this storm, see S. D. Burt and D. A. Mansfield, "The Great Storm of 15–16 October 1987," *Weather* 43 (1988): 90–110. Your author's house, at the time in rural Kent in southeastern England, was one of those affected. What struck me at the time was the number of neighboring houses that remained unrepaired for months afterward, suggesting that they were uninsured. For a catalog of other extreme European windstorms see J. E. Roberts et al., "The XWS Open Access Catalogue of Extreme European Windstorms from 1979 to 2012," *Natural Hazards and Earth System Sciences* 14 (2014): 2487–2501, http://centaur.reading.ac.uk/38441/.

19. T. M. Kostigen, "Government Lists 2013's Most Extreme Weather Events: 6 Takeaways" [online], *National Geographic*, January 22, 2014, http://news.nationalgeographic.com/news/2014/01/140122-noaa-extreme-weather-2013-climate-change-drought/. The six major events described are (1) worsening drought conditions in California; (2) Typhoon Haiyan in the Philippines (the strongest typhoon on record to hit land, it caused 5,700 deaths and spawned winds of 195 miles per hour); (3) the warmest year on record in Australia; (4) heavy rains in China and Russia (140 Russian towns reported the worst flooding in 120 years); (5) the coldest spring in the United Kingdom since 1962; and (6) the melting of Arctic and Antarctic ice much more than usual (the sixth and second lowest ice levels on record, respectively). For a popular account of Arctic ice levels, see also H. Briggs, "Arctic Sea Ice Hits Record Low," BBC News, March 21, 2015.

20. P. Miller, "Extreme Weather," *National Geographic*, September 2012. Miller's report includes an account of flash flooding in Tennessee in May 2010 due to unusually heavy downpours. ("We've got a building running into cars.") The dangers of the coming climate shift are made clear in J. Hansen, *Storms of My Grandchildren: The Truth About the Coming Climate Catastrophe and Our Last Chance to Save Humanity* (New York: Bloomsbury, 2010); and J. Hansen et al., "Ice Melt, Sea Level Rise and Superstorms: Evidence from Paleoclimate Data, Climate Modeling, and Modern Observations that 2°C Global Warming Is Highly Dangerous," *Atmospheric Chemistry and Physics* 15 (2015): 20059–20179.

21. J. Carey, "Storm Warnings: Extreme Weather Is a Product of Climate Change" [online], *Scientific American*, June 28, 2011, http://www.scientificamerican.com/article/extreme-weather-caused-by-climate-change/.

22. For the technical report of this climate attribution study, see S. C. Herring et al., eds., *Explaining Extreme Events of 2013 from a Climate Perspective*, Special supplement, *Bulletin of the American Meteorological Society* 95, no. 9 (2014); for a popular account of this study that reaches a different conclusion from that of its authors, see, for example, J. Taylor, "NOAA Report Destroys Global Warming Link to Extreme Weather," *Forbes*, October 9, 2014.

10. The World of Weather Forecasting

1. The timing of the Normandy landings (D-Day) was strongly dependent on weather conditions; the later German Ardennes offensive (Battle of the Bulge) was timed so that cloud cover would prevent Allied warplanes from operating effectively; NATO air operations over Serbia in the 1990s were compromised by bad weather.

2. In cold weather, people don't go to the stores much; in hot weather, they prefer to do things other than shopping. The price of orange juice rose 14% in 2006 after Hurricane Charley hit Florida in August 2004. See A. Martin, "Nature Getting the Blame for Costly Orange Juice," *New York Times*, December 2, 2006; and C. D. Ahrens, *Meteorology Today: An Introduction to Weather, Climate, and the Environment*, 10th ed. (Belmont, Calif.: Brooks/Cole, 2012).

3. Met Office, "How Accurate Are Our Public Forecasts?" May 24, 2016, www.metoffice.gov.uk/about-us/who/accuracy/forecasts.

4. Quoted in M. Chown, *What a Wonderful World: One Man's Attempt to Explain the Big Stuff* (London: Faber and Faber, 2013). Patrick Young is a financial analyst. Forecast accuracy is improving, but a 14-day weather forecast is only a little better than climatology—recall how chaos kicks in.

5. Quoted in M. Wall, "Weather Report: Forecasts Improving as Climate Gets Wilder," BBC News, September 25, 2014.

6. Funding will flow from the Weather Forecast Improvement Act (2013) and from the Disaster Relief Appropriations Act (2013). For more details on funding of severe-weather research and anticipated improvements in weather forecasting, see H. Christensen, "Banking on Better Forecasts: The New Maths of Weather Prediction," *Guardian*, January 8, 2015; J. Lubchenco and J. Hayes, "New Technology Allows Better Extreme Weather Forecasts," *Scientific American*, May 1, 2012, http://www.scientificamerican.com/article/a-better-eye-on-the-storm/; "Game-Changing Improvements in the Works for U.S. Weather Prediction," *Washington Post*, May 15, 2013; and Wall, "Weather Report." In what is seen as another recent high-profile embarrassment for the U.S. meteorological research community, the United States Air Force has recently switched from an American weather model to a British one. See J. Samenow, "Air Force's Plan to Drop U.S. Forecast System for U.K. Model Draws Criticism," *Washington Post*, April 20, 2015.

7. This interaction between very different length scales is typical of meteorology but is in fact not all that common in the wider field of physics. For example, quantum mechanics applies at very small length scales but is hardly noticeable at everyday length scales, where we use classical mechanics to describe nature. Similarly, classical Newtonian gravity works very well on the length scale of our solar system; we need invoke Einstein's general theory of relativity only for much larger length scales and much bigger gravitational systems.

8. Here is a simple mathematical analogy: persistence amounts to knowing the value of a function, whereas trends requires knowing the value of both the function and its first derivative.

9. For more on nowcasting, see C. Mass, "Nowcasting: The Next Revolution in Weather Prediction," 2011, http://www.atmos.washington.edu/~cliff/BAMSNowcast7.11.pdf.

10. Snowfall forecasting is performed via a number of rules of thumb that meteorologists have developed over the decades: "synoptic climatology methods," "Cook method," "Garcia method," and "magic dart."

11. There is much about practical weather forecasting, including rules of thumb and the presentation of forecasts—to which we now turn—in Ahrens, *Meteorology Today*; C. D. Ahrens, *Essentials of Meteorology: An Invitation to the Atmosphere* (Belmont, Calif.: Brooks/Cole, 2014); K. C. Harper, "Weather Forecasting by Numerical Methods," in *Discoveries in Modern Science: Exploration, Invention, Technology*, ed. J. Trefil (Farmington Mills, Mich.: Macmillan, 2015), 1214–1218; and H. M. Mogil, *Extreme Weather* (New York: Black Dog & Leventhal, 2007).

12. For example, the World Meteorological Organization has published *Guidelines on Graphical Presentation of Public Weather Services Products*, WMO/TD No. 1080 (Geneva: World Meteorological Organization, 2001).

13. The main motivation for these early British forecasts was maritime safety; 7,201 lives were lost off the coast of Britain between 1855 and 1860, many of which could have been saved given forewarning of bad weather, in the opinion of Robert Fitzroy, the Royal Navy admiral who founded the Met Office in 1860. He is the same Fitzroy who, many years earlier as captain of HMS *Beagle*, sailed Charles Darwin around the world on his epic voyage that would lead eventually to Darwin's theory of evolution. See P. Moore, "The Birth of the Weather Forecast," BBC News, April 30, 2015.

14. Pity poor Michael Fish, a Met Office spokesman who presented the weather on BBC television for many years and who made exactly this blooper immediately before the exceptionally severe windstorm over southern England in 1987—the worst such storm for three centuries. He has never lived it down.

15. See, for example, N. Silver, *The Signal and the Noise: Why So Many Predictions Fail* (New York: Penguin, 2012).

16. I do not necessarily mean active researchers who have become icons of their field, such as Albert Einstein and Stephen Hawking in theoretical physics, but the presenters of the subject. In physics, this might be Neil deGrasse Tyson or Michio Kaku in the United States and Brian Cox in the United Kingdom; in evolutionary biology, Stephen Jay Gould and Richard Dawkins; in ecology, Edward Wilson. These are prominent individuals, but hardly an army.

17. Quoted in G. Quill, "Many TV Weather Forecasters Lack Qualifications," *Toronto Star*, December 20, 2010.

18. Weather presenter Ericka Pino made the mistake of wearing a green dress to work, and became the unfortunate victim of a funny, if undignified, green-screen practical joke from co-workers, which can be seen as "Hilarious Green Screen Prank on Weather Girl," March 3, 2012, YouTube, https://www.youtube.com/watch?v=AzKB0OV2oqo.

19. Matt Watson of Bristol University, quoted in D. Shukman, "Geo-Engineering: Climate Fixes 'Could Harm Billions,'" BBC News, November 26, 2014.

20. See, for example, G. Dyer, *Climate Wars: The Fight for Survival as the World Overheats* (London: Oneworld, 2008).

21. An increase in weather modification research was called for in "Change in the Weather" [editorial], *Nature*, June 19, 2008, 957–958. While recognizing the patchy success of such programs in the past, it argues that current needs call for a more determined effort. See also "Fears of a Bright Planet" [editorial], *Economist*, December 13, 2014.

22. For more details on the Chinese weather modification program, see M. Williams, "Weather Engineering in China," *MIT Technological Review*, March 25, 2008.

23. For accessible accounts of both sides of the geoengineering debate, see C. Hamilton, "Geoengineering: Our Last Hope, or a False Promise?" *New York Times*, May 26, 2013; B. Lomberg, "Geoengineering: A Quick, Clean Fix?" *Time*, November 14, 2010; Shukman, "Geo-Engineering"; J. Vidal, "Geoengineering: Green versus Greed in the Race to Cool the Planet," *Guardian*, July 9, 2012, and "Geoengineering Side Effects Could Be Potentially Disastrous, Research Shows," *Guardian*, February 25, 2014; and M. Watson, "Why We'd Be Mad to Rule Out Climate Engineering," *Guardian*, October 8, 2013.

And That Wraps Up Your Weather for Today

1. Some climatologists and atmospheric scientists have defined an *Anthropocene Age*—a geological epoch that started within the past few hundred years (there is disagreement about the exact date). The idea is that human alterations of the planet are, from this date forward, so profound that a geologist with no prior

knowledge of Earth could visit the planet long after we are gone—say, in 1 million years—and detect our presence by means of stratigraphy or other geological analyses alone. See, for example, P. J. Crutzen and E. F. Stoerner, "The 'Anthropocene,'" *Global Change Newsletter* 41 (2000): 17–18; and W. F. Ruddiman, "The Anthropocene," *Annual Review of Earth and Planetary Sciences* 41 (2013): 45–68.

Bibliography

Ahrens, C. D. *Essentials of Meteorology: An Invitation to the Atmosphere*. Belmont, Calif.: Brooks/Cole, 2014.

——. *Meteorology Today: An Introduction to Weather, Climate, and the Environment*. 10th ed. Belmont, Calif.: Brooks/Cole, 2012.

Ahrens, C. D., and P. Samson. *Extreme Weather and Climate*. Belmont, Calif.: Brooks/Cole, 2011.

Akita, M., S. Yoshida, Y. Nakamura, T. Morimoto, T. Ushio, Z. Kawasaki, and D. Wang. "Effects of Charge Distribution in Thunderstorms on Lightning Propagation Paths in Darwin, Australia." *Journal of the Atmospheric Sciences* 68 (2011): 719–726.

Andrews, D. F., and A. M. Herzberg. *Data: A Collection of Problems from Many Fields for the Student and Research Worker*. New York: Springer, 1985.

Arsenault, C. "Risk of Water Wars Rises with Scarcity." Al Jazeera, August 26, 2012.

Baker, D. J. "Climate Change." In *Discoveries in Modern Science: Exploration, Invention, Technology*, edited by J. Trefil, 188–189. Farmington Mills, Mich.: Macmillan, 2015.

Barnaby, W. "Do Nations Go to War over Water?" *Nature*, March 19, 2009, 282–283.

Betts, A. K., and J. H. Ball. "Albedo over the Boreal Forest." *Journal of Geophysical Research* 102 (1997): 28901–28910.

Beven, J. L., L. A. Avila, E. S. Blake, D. P. Brown, J. L. Franklin, R. D. Knabb, R. J. Pasch, J. R. Rhome, and S. R. Stewart. "Atlantic Hurricane Season of 2005." *Monthly Weather Review* 136 (2008): 1109–1173.

Blake, E. S., and E. J. Gibney. "The Deadliest, Costliest and Most Intense United States Tropical Cyclones from 1851 to 2010 (and Other Frequently Requested Hurricane Facts)." NOAA Technical Memorandum NWS NHC-6. National Weather Service, National Hurricane Center, Miami, Fla., 2011.

Bluestein, H. B. "More Observations of Small Funnel Clouds and Other Tubular Clouds." *Monthly Weather Review* 133 (2005): 3714–3720.

Bouvier, A., and M. Wadhwa. "The Age of the Solar System Redefined by the Oldest Pb-Pb Age of a Meteoritic Inclusion." *Nature Geoscience* 3 (2010): 637–641.

Bowes, P. "California Drought: Will the Golden State Turn Brown?" BBC News, April 6, 2015.

Briggs, H. "Arctic Sea Ice Hits Record Low." BBC News, March 21, 2015.

Broecker, W. *The Great Ocean Conveyor: Discovering the Trigger for Abrupt Climate Change*. Princeton, N.J.: Princeton University Press, 2010.

Brooks, H. E., C. A. Doswell III, and R. B. Wilhelmson. "The Role of Midtropospheric Winds in the Evolution and Maintenance of Low-Level Mesocyclones." *Monthly Weather Review* 122 (1994): 126–136.

Brücker, G. "Vulnerable Populations: Lessons Learnt from the Summer 2003 Heat Wave in Europe." *Eurosurveillance* 10 (2005).

Budyko, M. "The Effect of Solar Radiation Variations on the Climate of the Earth." *Tellus A* 21 (1969): 611–619.

Burgess, P. "Variation in Light Intensity at Different Latitudes and Seasons, Effects of Cloud Cover, and the Amounts of Direct and Diffused Light." Paper presented at Continuous Cover Forestry Group Scientific Meeting, Westonbirt Arboretum, September 29, 2009. http://www.ccfg.org.uk/conferences/downloads/P_Burgess.pdf.

Burt, S. D., and D. A. Mansfield. "The Great Storm of 15–16 October 1987." *Weather* 43 (1988): 90–110.

Caldwell, M. "Unravelling Our Atmosphere." *Physics World* 27 (2014): 36–40.

Carey, J. "Storm Warnings: Extreme Weather Is a Product of Climate Change" [online]. *Scientific American*, June 28, 2011. http://www.scientificamerican.com/article/extreme-weather-caused-by-climate-change/.

Carrington, D. "Extreme Weather Becoming More Common, Study Says." *Guardian*, August 11, 2014.

"Change in the Weather" [editorial]. *Nature*, June 19, 2008, 957–958.

Chown, M. *What a Wonderful World: One Man's Attempt to Explain the Big Stuff*. London: Faber and Faber, 2013.

Christensen, H. "Banking on Better Forecasts: The New Maths of Weather Prediction." *Guardian*, January 8, 2015.

Coumou, D., V. Petoukhov, S. Rahmstorf, S. Petri, and H. J. Schellnhuber. "Quasi-resonant Circulation Regimes and Hemispheric Synchronization of Extreme Weather in Boreal Summer." *Proceedings of the National Academy of Sciences of the USA* 111 (2014): 12331–12336.

Coumou, D., and A. Robinson. "Historic and Future Increase in the Global Land Area Affected by Monthly Heat Extremes." *Environmental Research Letters* 8 (2013): 034018. DOI: 10.1088/1748-9326/8/3/034018.

Cowan, R. "Snowflake Growth Successfully Modeled from Physical Laws" [online]. *Scientific American*, March 16, 2012. http://www.scientificamerican.com/article/how-do-snowflakes-form/.

Cox, J. D. *Weather for Dummies*. Indianapolis: Wiley, 2000.

Crutzen, P. J., and E. F. Stoerner. "The 'Anthropocene.'" *Global Change Newsletter* 41 (2000): 17–18.

Dell'Amore, C. "'Snowball Earth' Confirmed: Ice Covered Equator" [online]. *National Geographic*, March 4, 2010. http://news.nationalgeographic.com/news/2010/03/100304-snowball-earth-ice-global-warming/.

Denny, M. *Blip, Ping, and Buzz: Making Sense of Radar and Sonar*. Baltimore: Johns Hopkins University Press, 2007.

——. *Lights On! The Science of Power Generation*. Baltimore: Johns Hopkins University Press, 2013.

——. *The Science of Navigation: From Dead Reckoning to GPS*. Baltimore: Johns Hopkins University Press, 2012.

——. "Weather Balloon Ascent Rate." *Physics Teacher* 54 (2016): 268–271.

Doesken, N. J. "Hail, Hail Hail! The Summertime Hazard of Eastern Colorado." *Colorado Climate* 17, no. 7 (1994). http://www.cocorahs.org/media/docs/hail_1994.pdf.

Doswell, C. A., III. "Historical Overview of Severe Convective Storms Research." *Electronic Journal of Severe Storms Meteorology* 2 (2007): 1–25.

Durre, I., R. S. Vose, and D. B. Wuertz. "Overview of the Integrated Global Radiosonde Archive." *Journal of Climate* 19 (2006): 53–68.

Dyer, G. *Climate Wars: The Fight for Survival as the World Overheats*. London: Oneworld, 2008.

Eakins, B. W., and G. F. Sharman. "Volumes of the World's Oceans from ETOPO1." NOAA National Geographic Data Center, Boulder, Colo., 2010.

Edwards, B. F., J. W. Wilder, and E. E. Scime. "Dynamics of Falling Raindrops." *European Journal of Physics* 22 (2001): 113–118.

Edwards, R. "Tropical Cyclone Tornadoes: A Review of Knowledge in Research and Prediction." *Electronic Journal of Severe Storms Meteorology* 7 (2012): 1–11.

Edwards, R., S. F. Corfidi, R. L. Thompson, J. S. Evans, J. P. Craven, J. P. Racy, and D. W. McCarthy. "Storm Prediction Center Forecasting Issues Relating to the 3 May 1999 Tornado Outbreak." *American Meteorological Society* 17 (2002): 544–558.

Falkowski, P., et al. "The Global Carbon Cycle: A Test of Our Knowledge of Earth as a System." *Science* 290 (2000): 291–296.

"Fears of a Bright Planet" [editorial]. *Economist*, December 13, 2014.

Flato, G., et al. "Evaluation of Climate Models." In *Climate Change 2013: The Physical Science Basis; Contribution of Working Group 1 to the Fifth Assessment Report of the Intergovernmental Panel on Climate Change*, edited by T. F. Stocker et al. Cambridge: Cambridge University Press, 2013.

Gat, J. R. *Isotope Hydrology: A Study of the Water Cycle*. London: Imperial College Press, 2010.

Gattuso, J.-P., et al. "Contrasting Futures for Ocean and Society from Different Anthropogenic CO_2 Emission Scenarios." *Science* 349 (2015): 45. DOI: 10.1126/science.aac4722.

Gettelman, A., M. L. Salby, and F. Sassi. "Distribution and Influence of Convection in the Tropical Tropopause Region." *Journal of Geophysical Research* 107 (2002): 4080. DOI:10.1029/2001JD001048.

Gleick, P. H., and M. Palaniappan. "Peak Water Limits to Freshwater Withdrawal and Use." *Proceedings of the National Academy of Sciences of the USA* 107 (2010): 11155–11162.

Goldberg, R. A. "A Review of Reported Relationships Linking Solar Variability to Weather and Climate." In *Solar Variability, Weather, and Climate*, edited by J. A. Eddy. Washington, D.C.: National Academies Press, 1982.

Goode, P. R., J. Qiu, V. Yurchyshyn, J. Hickey, M.-C. Chu, E. Kolbe, C. T. Brown, and S. E. Koonin. "Earthshine Observations of the Earth's Reflectance." *Geophysical Research Letters* 28 (2001): 1671–1674.

Grow, R. "Record Blocking Patterns Fueling Extreme Weather: Detailed Look at Why It's So Cold." *Washington Post*, March 21, 2013.

Guemas, V., F. J. Doblas-Reyes, I. Andreu-Burillo, and M. Asif. "Retrospective Prediction of the Global Warming Slowdown in the Past Decade." *Nature Climate Change* 3 (2013): 649–653.

Häkkinen, S., P. B. Rhines, and D. S. Worthen. "Atmospheric Blocking and Atlantic Multidecadal Ocean Variability." *Science* 334 (2011): 655–659.

Hamblyn, R. *The Invention of Clouds: How an Amateur Meteorologist Forged the Language of the Skies*. New York: Picador, 2001.

Hamilton, C. "Geoengineering: Our Last Hope, or a False Promise?" *New York Times*, May 26, 2013.

Hansen, J. *Storms of My Grandchildren: The Truth About the Coming Climate Catastrophe and Our Last Chance to Save Humanity*. New York: Bloomsbury, 2010.

Hansen, J., et al. "Ice Melt, Sea Level Rise and Superstorms: Evidence from Paleoclimate Data, Climate Modeling, and Modern Observations That 2°C Global Warming Is Highly Dangerous." *Atmospheric Chemistry and Physics* 15 (2015): 20059–20179.

Harper, K. C. "Weather Forecasting by Numerical Methods." In *Discoveries in Modern Science: Exploration, Invention, Technology*, edited by J. Trefil, 1214–1218. Farmington Mills, Mich.: Macmillan, 2015.

Harrabin, R. "Risk from Extreme Weather Rises." BBC News, November 26, 2013.

Harris, S. "Water Wars." *Foreign Policy*, September 18, 2014.

Hays, J. D., J. Imbrie, and N. J. Shackleton. "Variations in the Earth's Orbit: Pacemaker of the Ice Age." *Science* 194 (1976): 1121–1132.

"Heart of the Matter" [editorial]. *Nature*, July 28, 2011, 423–424.

Herring, S. C., M. P. Hoerling, T. C. Peterson, and P. A. Scott, eds. *Explaining Extreme Events of 2013 from a Climate Perspective*. Special supplement, *Bulletin of the American Meteorological Society* 95, no. 9 (2014).

Hillel, D. *Environmental Soil Physics: Fundamentals, Applications, and Environmental Considerations*. London: Academic Press, 1998.

Hoag, H. "Russian Summer Tops 'Universal' Heatwave Index." *Nature*, October 29, 2014.

Hocker, J. E., and J. B. Basara. "A Geographical Information Systems-Based Analysis of Supercells Across Oklahoma from 1994 to 2003." *Journal of Applied Meteorology and Climatology* 47 (2007): 1518–1538.

Holpuch, A. "Drought-Stricken California Only Has One Year of Water Left, NASA Scientist Warns." *Guardian*, March 16, 2015.

Houghton, J. T., Y. Ding, D. J. Griggs, M. Noguer, P. J. van der Linden, X. Dai, K. Maskell, and C. A. Johnson, eds. *Climate Change 2001: The Scientific Basis; Contribution of Working Group 1 to the Third Assessment Report of the Intergovernmental Panel on Climate Change.* Cambridge: Cambridge University Press, 2001.

House, A. "The Top 10 Wettest Places on Earth." *Daily Telegraph*, August 18, 2014.

Hu, Y., and Q. Fu. "Observed Poleward Expansion of the Hadley Circulation Since 1979." *Atmospheric Chemistry and Physics* 7 (2007): 5229–5236.

Huang, J., and M. B. McElroy. "Contributions of the Hadley and Ferrel Circulations to the Energetics of the Atmosphere over the Past 32 Years." *Journal of Climate* 27 (2014): 2656–2666.

Huang, S. P., H. N. Pollack, and P.-Y. Shen. "A Late Quaternary Climate Reconstruction Based on Borehole Heat Flux Data, Borehole Temperature Data, and the Instrumental Record." *Geophysical Research Letters* 35 (2008): L13703.

Iqbal, M. *An Introduction to Solar Radiation*. New York: Academic Press, 1983.

Karoff, C., and H. Svensmark. "How Did the Sun Affect the Climate When Life Evolved on the Earth?" arXiv (2010): 1003.6043.

Kasatkina, E. A., O. I. Shumilov, and M. Krapiec. "On Periodicities in Long Term Climate Variations near 68°N, 30°E." *Advances in Geoscience* 13 (2007): 25–29.

Kenney, W. L., D. H. Craighead, and L. M. Alexander. "Heat Waves, Aging, and Human Cardiovascular Health." *Medicine and Science in Sports and Exercise* 46 (2014): 1891–1899.

Kerr, R. A. "A Slowing Cog in the North Atlantic Ocean's Climate Machine." *Science* 304 (2004): 371–372.

Klinenberg, E. *Heat Wave: A Social Autopsy of Disaster in Chicago*. Chicago: University of Chicago Press, 2002.

Kostigen, T. M. "Government Lists 2013's Most Extreme Weather Events: 6 Takeaways" [online]. *National Geographic*, January 22, 2014. http://news.nationalgeographic.com/news/2014/01/140122-noaa-extreme-weather-2013-climate-change-drought/.

Lal, R., ed. *Encyclopedia of Soil Science*, 2nd ed. Boca Raton, Fla.: CRC Press, 2006.

Larsen, L. B., et al. "New Ice Core Evidence for a Volcanic Cause of the A.D. 536 Dust Veil." *Geophysical Research Letters* 35 (2008): L04708.

Laskar, J., A. Fienga, M. Gastineau, and H. Manche. "La2010: A New Orbital Solution for the Long-Term Motion of the Earth." *Astronomy and Astrophysics* 532 (2011): A89.

Leclercq, P. W., and J. Oerlemans. "Global and Hemispheric Temperature Reconstruction from Glacier Length Fluctuations." *Climate Dynamics* 38 (2012): 1065–1079.

Levitus, S., et al. "World Ocean Heat Content and Thermosteric Sea Level Change (0–2000 m), 1955–2010." *Geophysical Research Letters* 39 (2012): L10603.

Livio, M. *Brilliant Blunders: From Darwin to Einstein; Colossal Mistakes by Great Scientists That Changed Our Understanding of Life and the Universe*. New York: Simon and Schuster, 2013.

Lomberg, B. "Geoengineering: A Quick, Clean Fix?" *Time*, November 14, 2010.

Lorenz, R. D., and M. J. Myers. "Dust Devil Hazard to Aviation: A Review of United States Air Accident Reports." *Journal of Meteorology* 30 (2005): 178–184.

Lubchenco, J., and J. Hayes. "New Technology Allows Better Extreme Weather Forecasts." *Scientific American*, May 1, 2012. http://www.scientificamerican.com/article/a-better-eye-on-the-storm/.

Mambretti, S. E., ed. *Flood Risk Assessment and Management*. Southampton, Eng.: WIT Press, 2012.

Marshall, C. "Global Flood Toll to Triple by 2030." BBC News, March 4, 2015.

Marshall, T. C., and M. Stolzenburg. 2001. "Voltages Inside and Just Above Thunderstorms." *Journal of Geophysical Research* 106:4745–4768.

Martin, A. "Nature Getting the Blame for Costly Orange Juice." *New York Times*, December 2, 2006.
Mason, B. J. "The Oceans as a Source of Cloud-Forming Nuclei." *Pure and Applied Geophysics* 36 (1957): 148–155.
Mass, C.. "Nowcasting: The Next Revolution in Weather Prediction." 2011. http://www.atmos.washington.edu/~cliff/BAMSNowcast7.11.pdf.
McGuffie, K., and A. Henderson-Sellers. *A Climate Model Primer*. Hoboken, N.J.: Wiley, 2013.
Miller, P. "Extreme Weather." *National Geographic*, September 2012.
Milman, O. "Heatwave Frequency 'Surpasses Levels Previously Predicted for 2030.'" *Guardian*, February 17, 2014.
Miyahara, H., and Y. Yokoyama. "Influence of the Schwabe/Hale Solar Cycles on Climate Change During the Maunder Minimum." *Proceedings of the International Astronomical Union* S264 (2010): 427–433.
Mogil, H. M. *Extreme Weather*. New York: Black Dog & Leventhal, 2007.
Moore, P. "The Birth of the Weather Forecast." BBC News, April 30, 2015.
Netburn, D. "Mystery of the 'Faint Young Sun Paradox' May Be Solved." *Los Angeles Times*, June 10, 2013.
Neuenschwander, D. E. "The Physics of Tornadoes." *SPS Observer*, Fall 2011, 2–17.
Nishikawa, T., S. Maruyama, and S. Sakai. "Radiative Heat Transfer Analysis Within Three-Dimensional Clouds Subjected to Solar and Sky Irradiation." *Journal of the Atmospheric Sciences* 61 (2004): 3125–3133.
Open University. *Open Circulation*. Milton Keynes, Eng.: Open University, 2001.
Palmer, T. N. "Climate Extremes and the Role of Dynamics." *Proceedings of the National Academy of Sciences of the USA* 110 (2013): 5281–5282.
Papadatou-Pastou, M., M. Martin, M. R. Munafò, and G. V. Jones. "Sex Differences in Left-Handedness: A Meta-analysis of 144 Studies." *Psychological Bulletin* 134 (2008): 677–699.
Penna, A. N., and J. S. Rivers. *Natural Disasters in a Global Environment*. Chichester, Eng.: Wiley-Blackwell, 2013.
Peplow, M. "Snowflakes Made Easy." *Nature*, December 31, 2014.
Perkins, S. "Tornado Alley, U.S.A." *Science News* 161 (2002): 296.
Perkins, S. E., and L. V. Alexander. "On the Measurement of Heat Waves." *Journal of Climate* 26 (2013): 4500–4517.
Pidwirny, M. *Understanding Physical Geography*. 3 parts. Kelowna, B.C.: Our Planet Earth, 2014.
Pietz, D. A. *Engineering the State: The Huai River and Reconstruction in Nationalist China, 1927–1937*. New York: Routledge, 2002.
Povich, E. S. "Drought Is Not Just a California Problem." *USA Today*, April 19, 2015.

Quill, G. "Many TV Weather Forecasters Lack Qualifications." *Toronto Star*, December 20, 2010.

Radke, L. F., J. A. Coakley, and M. D. King. "Direct and Remote Sensing Observations of the Effects of Ships on Clouds." *Science* 246 (1989): 1146–1149.

Rahmstorf, S. "Rapid Climate Transitions in a Coupled Ocean–Atmosphere Model." *Nature*, November 3, 1994, 82–85.

Rangno, A. "Classification of Clouds." In *Discoveries in Modern Science: Exploration, Invention, Technology*, edited by J. Trefil, 199–201. Farmington Mills, Mich.: Macmillan, 2015.

Raspopov, O. M., V. A. Dergachev, J. Esper, O. V. Kozyreva, D. Frank, M. Ogurtsov, T. Kolström, and X. Shao. "The Influence of the de Vries (~200-year) Solar Cycle on Climate Variations: Results from the Central Asian Mountains and Their Global Link." *Paleogeography, Paleoclimatology, Paleoecology* 259 (2008): 6–16.

Reardon, S. "Titan Holds Clue to Faint Young Sun Paradox." *New Scientist*, January 2013.

Rensselaer Polytechnic Institute. "Argon Conclusion: Researchers Reassess Theories on Formation of Earth's Atmosphere." *ScienceDaily*, September 24, 2007. www.sciencedaily.com/releases/2007/09/070919131757.htm.

Rex, D. F., ed. *Climate of the Free Atmosphere*. Vol. 4 of *World Survey of Climatology*. New York: Elsevier, 1969.

Rice, D. "Killer Cold: Winter Is Deadlier Than Summer in U.S." *USA Today*, July 30, 2014.

Roberts, J. F., A. J. Champion, L. C. Dawkins, K. I. Hodges, L. C. Shaffrey, D. B. Stephenson, M. A. Stringer, H. E. Thornton, and B. D. Youngman. "The XWS Open Access Catalogue of Extreme European Windstorms from 1979 to 2012." *Natural Hazards and Earth System Sciences* 14 (2014): 2487–2501.

Robine, J.-M., S. L. K. Cheung, S. Le Roy, H. Van Oyen, C. Griffiths, J.-P. Michel, and F. R. Herrmann. "Death Toll Exceeded 70,000 in Europe During the Summer of 2003." *Comptes Rendus Biologies* 331 (2008): 171–178.

Roble, R. G., and I. Tzur. "The Global Atmospheric-Electric Circuit." In *The Earth's Electrical Environment*, edited by National Research Council, 206–231. Washington, D.C.: National Academies Press, 1986.

Rosenfeld, D., and I. M. Lensky. "Satellite-Based Insights into Precipitation Formation Processes in Continental and Maritime Convective Clouds." *Bulletin of the American Meteorological Society* 79 (1998): 2457–2476.

Ross, D. *Introduction to Oceanography*. New York: HarperCollins, 1995.

Ruddiman, W. F. "The Anthropocene." *Annual Review of Earth and Planetary Sciences* 41 (2013): 45–68.

Russo, S., A. Dosio, R. G. Graversen, J. Sillmann, H. Carrao, M. B. Dunbar, A. Singleton, P. Montagna, P. Barbola, and J. V. Vogt. "Magnitude of Extreme

Heat Waves in Present Climate and Their Projection in a Warming World." *Journal of Geophysical Research* 119 (2014): 12500–12512.

Ruth, D. P. "Interactive Forecast Preparation—The Future Has Come." In *Proceedings of the Interactive Symposium on the Advanced Weather Interactive Processing System (AWIPS)*, Orlando, Fla., January 13–17, 2002, American Meteorological Society, 20–22.

Rybach, L. "Geothermal Sustainability." *Geo-Heat Centre Quarterly Bulletin* 28 (2007): 2–7.

Samenow, J. "Air Force's Plan to Drop U.S. Forecast System for U.K. Model Draws Criticism." *Washington Post*, April 20, 2015.

——. "Game-Changing Improvements in the Works for U.S. Weather Prediction." *Washington Post*, May 15, 2013.

Schatten, K. H., and A. S. Endal. "The Faint Young Sun-Climate Paradox: Volcanic Influences." *Geophysical Research Letters* 9 (1982): 1309–1311.

Schneider, T. "The General Circulation of the Atmosphere." *Annual Review of Earth and Planetary Science* 34 (2006): 655–688.

Seager, R. "The Source of Europe's Mild Climate." *American Scientist* 94 (2006): 334–341.

Seager, R., D. S. Battisti, J. Yin, N. Gordon, N. Naik, A. C. Clement, and M. A. Cane. "Is the Gulf Stream Responsible for Europe's Mild Winters?" *Quarterly Journal of the Royal Meteorological Society* 128 (2002): 2563–2586.

Shukman, D. "Geo-Engineering: Climate Fixes 'Could Harm Billions.'" BBC News, November 26, 2014.

Silver, N. *The Signal and the Noise: Why So Many Predictions Fail.* New York: Penguin, 2012.

Sinclair, P. C. "Some Preliminary Dust Devil Measurements." *Monthly Weather Review* 92 (1964): 363–367.

Sircombe, K. "Rutherford's Time Bomb." *New Zealand Herald* (Auckland), May 15, 2004.

Smith, J. A. E. Hui, M. Steiner, M. L. Baeck, W. F. Krajewski, and A. A. Ntelekos. "Variability of Rainfall Rate and Raindrop Size Distributions in Heavy Rain." *Water Resources Research* 45 (2009): WO4430. DOI: 10.1029/2008/WR006840.

Smith, R. "Hurricane Force." *Physics World*, June 2006, 32–37.

Soden B. J., and I. M. Held. "An Assessment of Climate Feedbacks in Coupled Ocean–Atmosphere Models," *Journal of Climate* 19 (2006) 3354–3360.

Solomon, S., D. Qin, M. Manning, Z. Chen, M. Marquis, K. B. Averyt, M. Tignor, and H. L. Miller, eds. *Climate Change 2007: The Scientific Basis; Contribution of Working Group 1 to the Fourth Assessment Report of the Intergovernmental Panel on Climate Change*. Cambridge: Cambridge University Press, 2007.

Sondergard, S. E. *Climate Balance: A Balanced and Realistic View of Climate Change*. Mustang, Okla.: Tate, 2009.

Strahler, A. H., and A. Strahler. *Physical Geography: Science and Systems of the Human Environment*. 2nd ed. New York: Wiley, 2003.

Sverdrup, K. A., A. C. Duxbury, and A. B. Duxbury. *Fundamentals of Oceanography*. 5th ed. New York: McGraw-Hill, 2006.

Taylor, J. "NOAA Report Destroys Global Warming Link to Extreme Weather." *Forbes*, October 9, 2014.

Trenberth, K. E., J. T. Fasullo, and J. Kiehl. "Earth's Global Energy Budget." *Bulletin of the American Meteorological Society* 90 (2009): 311–323.

Trenberth, K. E., and T. J. Hoar. "The 1990–1995 El Niño–Southern Oscillation Event: Longest on Record." *Geophysical Research Letters* 23 (1996): 57–60.

Trenberth, K. E., et al. "Observations: Surface and Atmospheric Climate Change." In *Climate Change 2007: The Physical Science Basis; Contribution of Working Group 1 to the Fourth Assessment Report of the Intergovernmental Panel on Climate Change*, edited by S. Solomon, D. Qin, M. Manning, Z. Chen, M. Marquis, K. B. Averyt, M. Tignor, and H. L. Miller. Cambridge: Cambridge University Press, 2007.

Vidal, J. "Geoengineering: Green versus Greed in the Race to Cool the Planet." *Guardian*, July 9, 2012.

——. Geoengineering Side Effects Could Be Potentially Disastrous, Research Shows." *Guardian*, February 25, 2014.

Wall, M. "Weather Report: Forecasts Improving as Climate Gets Wilder." BBC News, September 25, 2014.

Wang, P. K. *Physics and Dynamics of Clouds and Precipitation*. Cambridge: Cambridge University Press, 2013.

Watson, M. "Why We'd Be Mad to Rule Out Climate Engineering." *Guardian*, October 8, 2013.

Weart, S. "General Circulation Models of the Atmosphere." In *Discoveries in Modern Science: Exploration, Invention, Technology*, edited by J. Trefil, 41–46. Farmington Mills, Mich.: Macmillan, 2015.

Weaver, A. *Generation Us: The Challenge of Global Warming*. Victoria, B.C.: Orca, 2011..

Whiton, R. C., P. L. Smith, S. G. Bigler, K. E. Wilk, and A. C. Harbuck. "History of Operational Use of Weather Radar by U.S. Weather Services, Part II: Development of Operational Doppler Weather Radars." *Weather and Forecasting* 13 (1988): 244–252.

Williams, J. "Doppler Radar Measures 318 mph Wind in Tornado." *USA Today*, May 17, 2005.

Williams, M. "Weather Engineering in China." *MIT Technological Review*, March 25, 2008.

Witze, A., and J. Kanipe. *Island on Fire: The Extraordinary Story of Laki, the Volcano That Turned Eighteenth-Century Europe Dark.* London: Profile Books, 2014.

Wood, G. D. *Tambora: The Eruption That Changed the World.* Princeton, N.J.: Princeton University Press, 2014.

Wordsworth, R., and R. Pierrehumbert. "Hydrogen-Nitrogen Greenhouse Warming in Earth's Early Atmosphere." *Science* 339 (2013): 64–67.

World Meteorological Organization. *Guidelines on Graphical Presentation of Public Weather Services Products.* WMO/TD No. 1080. Geneva: World Meteorological Organization, 2001.

Wyett, R. E. "Pressure Drop in a Dust Devil." *Monthly Weather Review* 82 (1954): 7–8.

Zängl G., and K. P. Hoinka. "The Tropopause in the Polar Regions." *Journal of Climate* 14 (2001): 3117–3139.

Index